P. L. (Peter Lund) Simmonds

Coffee and Chicory

Their Culture, Chemical Composition, Preparation for Market and

Consumption

P. L. (Peter Lund) Simmonds

Coffee and Chicory
Their Culture, Chemical Composition, Preparation for Market and Consumption

ISBN/EAN: 9783743687110

Printed in Europe, USA, Canada, Australia, Japan

Cover: Foto ©berggeist007 / pixelio.de

More available books at **www.hansebooks.com**

COFFEE AND CHICORY

CULTURE, CHEMICAL COMPOSITION, PREPARATION
FOR MARKET, AND CONSUMPTION,

WITH

SIMPLE TESTS FOR DETECTING ADULTERATION,

AND

PRACTICAL HINTS FOR THE PRODUCER
AND CONSUMER.

BY

P. L. SIMMONDS,

AUTHOR OF "THE COMMERCIAL PRODUCTS OF THE VEGETABLE KINGDOM,"
"A DICTIONARY OF TRADE PRODUCTS," &c.

WITH NUMEROUS ILLUSTRATIONS.

LONDON:

E. & F. N. SPON, 16, BUCKLERSBURY.

1864.

PREFACE.

A PRACTICAL essay on the culture and preparation of
coffee for market in the various producing countries of
the world, brought down to the present time, has long
been wanted, especially as the sources of supply have
changed so much of late years. Porter's "Tropical Agri-
culturist" has long been out of print, and my own work
on "The Commercial Products of the Vegetable Kingdom"
is too expensive and too diffuse for ordinary reference.
The present hand-book deals with the subject in a popular
form, but, at the same time, supplies correct information
on most points, combined with the fullest descriptive and
statistical details respecting every coffee-producing country.
For much of the information relating to coffee cultivation
in Ceylon, I am indebted to a small treatise by Mr. G. C.
Lewis, privately published in that island. For the views
of buildings and scenery, I am under obligations to Sir

Emerson Tennent and Messrs. Worms, who kindly lent me original drawings and photographs—whilst the microscopic representations of pure and adulterated coffee and chicory are copied, by permission, from Dr. Hassall's elaborate work on "Food and its Adulterations." Trusting that this little work may be found useful and interesting to a large class, I send it forth as the pioneer of other hand-books on the great staples of commerce.

P. L. S.

8, Winchester-street, S.W.,
July, 1864.

CONTENTS.

COFFEE.

b

SECTION X.

CHICORY.

SECTION I.

COFFEE AND CHICORY.

COFFEE.

SECTION I.

BOTANICAL DESCRIPTION.

THE coffee-tree—*Coffea arabica*, Linn.—is a plant belonging to the natural order *Cinchonaceæ*. It is a large erect bush, quite smooth in every part; leaves oblong lanceolate, acuminate, shining on the upper side, wavy, deep green above, paler below; stipules subulate, undivided. Peduncles axillary, short, clustered; corollas white, funnel-shaped, sweet-scented, with four or five oblong-spreading twisted lobes. Fruit a compressed drupe, furrowed along the side, crowned by the calyx. Seeds solitary, plano-convex, with a deep furrow along the flat side. Putamen like parchment.

The generic name given to the plant by Linnæus was taken, it is said, from Coffee, a province of Narea, in Africa where it grows in abundance.

Plate 1 represents a branch of the coffee-tree in blossom

B

and fruit, and the lettered figures at the foot have reference to the dissection of the flower and fruit.

A—The flower, cut open, to show the situation of the five filaments, with their summits lying upon them.

B—Represents the flower cup, with its four small indentations enclosing the germ or embryo seed-vessel, from the middle of which arises the style, terminated by the two reflexed spongy tops.

C—The fruit entire, marked at the top with a puncture like a navel.

D—The fruit open, to show that it consists ordinarily of two seeds, which are surrounded by the pulp.

E—The fruit cut horizontally, to show the seeds as they are placed erect, with their flat sides, together.

F—One of the seeds taken out, with the membrane or parchment upon it.

G—The same with the parchment torn open, to give a view of the seed.

H—The seed without the parchment.

Lindley and Paxton only enumerate two species: *C. arabica*, native of Yemen, and *C. paniculata*, indigenous to Guiana.

Continental botanists, however, describe no less than eight other species: four inhabiting Peru, *C. microcarpa*, *C. umbellata*, *C. acuminata*, and *C. subsessilis*; two indigenous to the West Coast of Africa, *C. laurina* and *C. racemosa*; and two natives of the East Indies, *C. bengalensis* and *C. Indica*. Some of these are probably mere varieties.

Whatever its origin may have been, there can be no doubt that there are three kinds or species now grown, differing materially from each other.

The Arabian or Mocha coffee is characterised by having a small and more brittle leaf, with branches shorter, and more upright than the Jamaica and Ceylon coffee; and by its berry being almost always, or at least very frequently, single seeded, and the seed cylindrical and plump.

The Jamaica coffee-tree has a larger and more pliable leaf, longer and more drooping branches, and berries almost always containing two seeds. (*The Ethiopian.*)

The great difference now existing between the two kinds, may possibly have originated in the change of soil, climate, and season, operating through a series of years; but this difference is so decided, and so strongly marked, that the veriest tyro can in a moment pronounce of either.

The East India or Bengal coffee-tree differs much from all others, but is in every respect a veritable coffee.

The leaf is smaller, and lighter green, than the foregoing variety; its berry is infinitely smaller, and when ripening, turns black instead of blood-red. Coffee made from it is of excellent flavour, and much liked.

Within the tropics, coffee thrives best at an elevation of 1200 to 3000 feet, and rarely grows above 6000 feet. It may be cultivated as far as 30° north latitude, where the mean temperature is about 70°.

In the western hemisphere coffee is grown in many of the West India Islands, in Central America, the northern republics of South America, Berbice, Cayenne, and Brazil. In Africa it is grown in Liberia and other parts of Western Africa, at St. Helena, in Egypt, and Mozambique, and a little in Natal. Passing eastward we find it in Arabia, one of the oldest seats of culture, the southern peninsula of India, Ceylon, Bourbon, Java, Célèbes, and other parts of the Eastern Archipelago, Siam, and some of the Pacific Islands.

Coffee-plants are able to bear an amount of cold which is little known or thought of. The high and cold regions of Jamaica near St. Catherine's Peak, and the foot of the Great Blue Mountain Peak, both situated at some 6000 feet above the level of the sea; and, again, the mountains of Arabia, the Neilgherries, and Ceylon, furnish instances of the great degree of cold that the coffee-plant will endure. More than this, it is an established fact that it bears a larger, plumper, and far more aromatic berry at these altitudes than in a lower situation and in a warmer temperature. The coffee produced on plantations near the foot of the Blue Mountain Peak, in Jamaica, is the finest in the world. In Arabia, likewise, the cold at night is sometimes intense; yet who will dispute the goodness of Mocha coffee?

Nothing can exceed the beauty of the rows or walks planted with coffee-trees, from their pyramidical shape and glossy dark leaves, amongst which are hanging the ripe, scarlet-coloured berries. A writer, in his "Impressions of the West Indies," thus speaks: "Anything in the way of cultivation more beautiful or more fragrant than a coffee-plantation I had not conceived, and oft did I say to myself that if ever I became, from health or otherwise, a cultivator of the soil within the tropics, I would cultivate the coffee-plant, even though I did so irrespective altogether of the profit that might be derived from so doing. Much has been written, and not without justice, of the rich fragrance of an orange-grove, and at home we ofttimes hear of the sweet odours of a bean-field. I have, too, often enjoyed, in the Carse of Stirling and elsewhere in Scotland, the balmy breezes as they swept over the latter, particularly when the sun had burst out with unusual strength after a shower of rain. I have likewise in Martinique, Santa Cruz, Jamaica, and Cuba, inhaled the breezes wafted from the orangeries, but not for a moment would I compare either with the exquisite aromatic odours from a coffee-plantation in full bloom, when the hill-side—covered over with regular rows of the shrubs, with their millions of jasmine-like flowers—showers down upon you as you ride up between the plants a perfume

Plate 2.—A Coffee Plantation in Jamaica.

of the most delicately delicious description. 'Tis worth going to the West Indies to see the sight and inhale the perfume."

Plate 2 represents a coffee plantation in Jamaica.

In the culture of the tree there is a singular difference in the western and eastern hemispheres, inasmuch as in the former shade is considered injurious, whilst in the latter it is held to be desirable, if not absolutely necessary.

SECTION II.

Coffee, although taking its common and specific names from Arabia, is not a native plant of that country, but of Abyssinia, where it is found both in the wild and cultivated state. From that country it was brought to Arabia, in comparatively very recent times. Mr. Lane states that it was first used there about the year 1450. It was not known to the Arabs, therefore, for more than eight hundred years after the time of Mahomed, and was introduced only between forty and fifty years before the discovery of America. The Arabians called coffee kăhwăh, which is an old word in their language for wine. The unlucky word gave rise to a dispute about the legality of its use among the Mahomedan doctors, who, mistaking the word for the thing it represented, denounced as a narcotic that which was anti-narcotic. They were beaten, and coffee has ever since become a legitimate and favourite potable of the Arabs.

In a century its use spread to Egypt and other parts of the Turkish empire. For two centuries from its introduction into Arabia, the use of coffee seems to have been confined to the Mahomedan nations of Western Asia; and, considering its rapid spread and popularity among the European nations, it is remarkable that it has not, like tobacco, extended to the Hindus, the Hindu-Chinese, the Japanese, or the tribes of the Indian Archipelago, who no more use it than the Europeans do the betel preparation. The high price of coffee and the low cost of tobacco, most likely afford the true solution of the difference. One striking result of the use of coffee first, and then of tobacco, among the Mahomedan nations is well deserving of notice. These commodities have been in a great measure substituted for wine and spirits, which had been largely, although clandestinely, used before, and hence a great improvement in the sobriety of Arabs, Persians, and Turks. I give this interesting fact on the authority of Mr. Lane, who mentions it in the notes to his translation of the Arabian Nights.*

From Turkey coffee found its way to Europe. It came in use in England before either tea or chocolate. A Turkey merchant of London, of the name of Edwards, is said to have brought the first bag of coffee to England, and his Greek servant to have made the first dish of English coffee about 1652. But it is stated in the Life of Wood, the antiquary, that "in 1651, one Jacob, a Jew, opened a coffee-house at the Angel, in the parish of St. Peter-in-the-East, Oxon; and there it was, by some who delighted in noveltie, drunk. When he left Oxon, he sold it in Old Southampton-buildings, in Holborne, near London, and was living there in 1671."

Coffee-houses were soon after opened in various parts of the metropolis, as also in other parts of the kingdom, for vending it. The excise officers visited the coffee-houses at fixed periods, and took an account of the number of gallons of the liquid that were made, upon which a duty of 4d. per gallon was charged until 1689.

Three years after the first introduction of coffee upon the statute books, the increase of houses for its sale had become so great, that by the Act passed in 1663, "For the better ordering and collecting the duty of excise, and preventing the abuses therein" (15 Chas. II., cap. 11, sect. 15), express provision is made for the licensing of all coffee-

---

* Mr. J. Crawfurd on the History of Coffee, in the Statistical Society's Journal, vol. xv. p. 51.

houses at the quarter sessions, under a penalty of 5*l.* for every month during which any person should retail coffee, chocolate, or tea, without having first procured such license from the magistrates. From that time to the revolution, coffee-houses multiplied so rapidly that, when Ray published his " History of Plants" in 1688, he estimated that the coffee-houses in London were at that time as numerous as in Cairo itself; whilst similar places of accommodation were to be met with in all the principal cities and towns in England.

There are now in London alone more than 1500 coffee-houses, besides confectioners' shops, and other places where coffee is vended.

For half a century at least Arabia furnished all the coffee that Europe consumed, which, therefore, must have been very trifling. It was, in fact, long the luxury of a few fashionable people, with whom, however, it must have been in general use sixty years after its introduction, as we find from the well-known passage of the " Rape of the Lock," published in 1712, in which politicians are described as seeing through it " with half-shut eyes."

Le Grand d'Aussy, in his "Vie Privée des Français," gives a curious and interesting account of the first introduction of the use of coffee in France. As early as 1658 some merchants of Marseilles introduced the use of coffee into that city, and Thévenot, after his return from his Eastern travels, about the year 1658, regaled his guests with coffee after dinner.

" This, however," says Le Grand, " was but the eccentricity of a traveller, which would not come into fashion among such a people as the Parisians. To bring coffee into credit, some extraordinary and striking circumstance was necessary. This circumstance occurred on the arrival, in 1669, of an embassy from the Grand Seigneur Mahomet IV.

to Louis XIV. Soliman Aga, chief of the mission, having passed six months in the capital, and during his stay having acquired the friendship of the Parisians by some traits of wit and gallantry, several persons of distinction, chiefly women, had the curiosity to visit him at his house. The manner in which he received them not only inspired a wish to renew the visit, but induced others to follow their example. He caused coffee to be served to his guests, according to the custom of his country; for since fashion had introduced the custom of serving this beverage among the Turks, civility demanded that it should be offered to visitors, as well as that those should not decline partaking of it. If a Frenchman, in a similar case, to please the ladies, had presented to them this black and bitter liquor, he would be rendered for ever ridiculous. But the beverage was served by a Turk—a gallant Turk—and this was sufficient to give it inestimable value. Besides, before the palate could judge, the eyes were seduced by the display of elegance and neatness which accompanied it, by those brilliant porcelain cups into which it was poured, by napkins with gold fringes, on which it was served to the ladies; add to this the furniture, the dresses, and the foreign customs, the strangeness of addressing the host through the interpreter, being seated on the ground, on tiles, &c., and you will allow that there was more than enough to turn the heads of French women. Leaving the hotel of the ambassador with an enthusiasm easily imagined, they hastened to their acquaintances, to speak of the coffee of which they had partaken; and Heaven only knows to what a degree they were excited."

The extravagant price of coffee, notwithstanding that the fashion of drinking it was established, prevented it from coming into general use. It was only to be had, according to Le Grand, at Marseilles, and even there not in any quantity. Labat, quoted by him, states that the price at this time was the enormous one of forty crowns a pound. In 1672, an Armenian opened in Paris the first coffee-house, on the plan of those he had seen at Constantinople. Pascal was followed by a crowd of imitators, whose numbers became so great in 1676, that it was found necessary to form them into a society by statute.

As to the European names of coffee; they are all, observes Mr. Crawfurd, from the same source, the old Arabic word for wine, kăhwăh, which is composed of a very guttural *k*, unpronounceable by Europeans, except by an awkward effort, of the labial *w*, and of two short vowels ă, with an aspirate at the end of each syllable. The Turks have changed the labial *w* into *v*, and the European nations, who took the word directly from them, have corrupted the word by converting the labial *v* into the labial *f*, by substituting an ordinary *k* or hard *c* for the Arabic guttural, by omitting both the aspirates, and by converting the last short ă into ĕ, or as with ourselves, always the greatest corruptors of orthography, changing both the vowels.

The Mahomedans distinguish three kinds of kăhwăh— wine, or anything that inebriates ; the extract from the pulp which contains the coffee-berry ; and that from the berry itself. The deep brown colour of the liquor occasioned its being called the syrup of the Indian mulberry, under which specious name it first became fashionable in Europe ; and some who imported the pulp called it the " flower of the coffee-tree," but it failed in use. Coffee is used in vast quantities by the Turks and Arabians, and with peculiar propriety, as it counteracts the narcotic effects of opium, to the use of which they are so much addicted.

The history of the cultivation of coffee by European nations in their colonies is singular. The old Dutch East India Company carried on some traffic with the Arabian ports in the Red Sea; and about the year 1690, the Dutch

Governor-General of India, Van Hoorne, caused some ripe coffee-seeds to be brought to Java; they were planted, grew, and produced fruit. He sent a single plant home from Batavia to Nicholas Witsen, the Governor of the East India Company, which arrived safe, was planted in the Botanic Gardens of Amsterdam, where it prospered, produced fruit, and the fruit young plants. From the Amsterdam garden plants were sent to the Dutch colony of Surinam, and the planters entered on the cultivation of coffee in 1718. The authority for this is the celebrated physician and botanist, Boerhaave, in his Index of the Leyden Garden. In ten years after its cultivation in Surinam it was introduced from that colony by the English into Jamaica. It was sent to Martinique from France in 1720. The first coffee-plant cultivated in Brazil, now the greatest producing country in the world, was reared by a Franciscan monk, of the name of Vellosa, in the garden of the convent of San Antonio, near Rio Janeiro; it throve, and the monk presented its ripe fruit to the viceroy, Lavrado. He judiciously distributed it to the planters, who commenced its cultivation in 1774. From Java the coffee-plant was conveyed to Sumatra, to Célèbes, to the Philippines, and in our own times to Malabar, Mysore, and Ceylon. The few coffee-berries brought from Mocha to Batavia are the parents of the vast quantity now produced, all the coffee now consumed (exceeding 500,000,000 lbs.), save the trifle yielded by Arabia, has the same origin, and the great cultivation and commerce in coffee has all sprung up in less than one hundred and seventy-five years.

PRODUCTION AND SUPPLY.

THE changes in the sources of supply of coffee within the last quarter of a century are very remarkable. The British possessions in the East, where land and labour is cheap, have taken the place which our Western possessions formerly occupied.

The British West India Islands and Demerara have fallen off in their production of coffee from 30,000,000 lbs. to 4,000,000 lbs. San Domingo, Cuba, and the French West India colonies are also gradually giving up coffee culture in favour of other staples. It is chiefly Brazil, some of the Central American republics, Java, Ceylon, and British India, that are able to render coffee a profitable crop.

At the close of the last century the consumption of coffee was under one million pounds; the only descriptions then known in the London market were Grenada, Jamaica, and Mocha—the two former averaging about 5l. per cwt., and the latter 20l. per cwt. Grenada coffee is now unknown, and Ceylon, Java, and Brazil are the largest producers.

In 1760 the total quantity of coffee consumed in the United Kingdom was 262,000 lbs., or three-quarters of an ounce to each person in the population.

From 1801 to 1804 the average quantity of coffee consumed by each individual of the population was only about 1 oz., whilst 1½ lbs. of tea per head was used. From 1805 to 1809 the consumption of coffee was 3 oz. per head. From 1810 to 1824, when the duty was reduced by about one-

third, the consumption was 8 oz. After this, when the duty
on British-grown coffee was further reduced to 9d. and 6d.
the pound, the consumption rose to 1 lb., and by 1850 to
1½ lbs. But this consumption was not uniform for the
United Kingdom, for while in England 1 lb. 12 oz. was
used, in Scotland only 6 oz. were consumed, and in Ireland
but 2 oz.

The quantities of coffee consumed in Great Britain in
each decennial period, comparing the consumption with the
growth of the population, and exhibiting the influence of
high and low duties, are shown by the following statement.
The figures up to 1841 are from Porter's " Progress of the
Nation." Those since are computed from official documents :

	No. of lbs. con-sumed.	Duty on B. P. coffee.		Population of Great Britain.	Average consumption.	
		s.	d.		lbs.	oz.
1801......	750,000	1	6	10,942,646	0	1·09
1811......	6,390,122	0	7	12,596,803	0	8·12
1821......	7,327,283	1	0	14,391,631	0	8·01
1831......	21,862,264	0	6	16,262,801	1	5·49
1841......	27,298,322	0	6	18,532,335	1	7·55
1851......	32,504,545	0	3	21,000,000	1	4·98
1861......	35,204,040	0	3	23,266,755	1	1·33

It appears from the foregoing figures, that, when the duty
amounted to 1s. 6d. per lb., the use of coffee was confined
altogether to the rich. The quantity then used throughout
the kingdom scarcely exceeded on the average one ounce
for each inhabitant in the year.

Although about a quarter of a century ago the average
consumption rose to nearly 1½ lb., it has since been gradually
declining, for last year (1863) the total consumption was,
with an increased population, 2¼ million pounds below the
quantity taken for consumption in 1861.

The following table shows the changes in our sources of

supply of coffee even in the last ten years, taking the quantities entered for consumption only.

COFFEE TAKEN FOR CONSUMPTION IN THE UNITED KINGDOM.

	1853. lbs.		1862. lbs.
British India (including Ceylon)...	24,980,375	...	5,422,369
Ceylon	—	...	23,886,007
British West Indies and Guiana...	2,742,913	...	2,380,683
Central America	4,948,848	...	2,087,638
Brazil	814,133	...	280,837
Venezuela	1,033,071	...	—
Hayti	862,254	...	20,701
Java	112,892	...	—
Holland, &c.	442,863	...	8,862
Egypt.................................	112,360	...	90,932
United States	112,673	...	30,476
New Granada	—	...	133,144
Chili	379,930	...	—
Mauritius	61,884	...	—
Portugal	—	...	23,052
Philippine Islands....................	—	...	82,820
Other parts	487,574	...	216,650
	37,091,770		34,664,135
Exported on drawback.........	108,648	...	212,369
	36,983,122		34,451,766

There is imported into Europe annually about 270,000,000 pounds of coffee, of which France consumes one-sixth, the consumption there having increased fully fifty per cent. within a very brief period.

SECTION IV.

COMMERCIAL VARIETIES OF COFFEE.

THE coffee-berry of Cayenne is rather convex, irregular, of a dull green, covered with a slight pellicle. It is analogous to Mocha, and of a pleasant aromatic flavour.

That of Guadaloupe is elongated, larger, of a dark greyish green, and nearly always without any pellicle.

The coffee-berry of Martinique is also large, oval, flat on one side, with the furrow deep and straight for the greater part of its length, but diverging at the ends. Its odour is agreeable, and the flavour strong when used alone, but it is generally mixed with Mocha.

The Mocha berry is very variable in form, size, and colour, but it is generally more round or compressed than other coffees; its odour is strong and agreeable, and very characteristic. Many of the seeds are often covered by the endocarp, while others are without the pellicle. A great number also are rounded, and the involuted edges form a deep furrow, differing from the ordinary one. The form of these seeds is due to the abortion of the other half of the fruit, which gives it this particular formation.

Aden, *alias* Mocha, coffee is, along with the other coffees of the Red Sea, sent first to Bombay in Arab ships, where it is "garbelled" (picked), previously to its being exported to England. The bean is always broad and small, and the climate of India is supposed to improve its flavour. The seed of the Berbera (Abyssinian) plant is usually called long-berried Mocha.

The Java and East Indian, next in quality, are larger, and of a paler yellow. The Ceylon berries are of irregular sizes, ill-shaped, and of a spotted dirty cream-colour. The terms "Plantation" and "Native" coffees, as applied to Ceylon berries, are distinctions arising from one being the cultivated coffee of the estates of the planters, which are better attended to and prepared for market, while the other is that grown in a wild or careless manner by the natives about their dwellings, and more rudely prepared. Java coffee is chiefly prized in the market for its delicacy of flavour, but in point of strength it falls short of the West Indian.

Of Bourbon coffee there are in commerce two qualities, fine and ordinary. The first is in small seeds, well selected for size, of a variable colour, yellow or green, with little pellicle, the furrow slightly indented, and it has a sweet odour. The second is badly assorted as regards form and colour, and its odour less agreeable.

The Jamaica coffee-berry is medium-sized, of a greenish blue colour, rather oblong, and smooth to the touch. It has a strong, agreeable smell, and excellent flavour, and when carefully picked and sorted, fetches about the highest price of any kind.

Porto Rico is a middle-sized coffee, of a pure and agreeable flavour; the colour of the better sorts is a bluish green; and of the common, yellow.

The West Indian and Brazilian coffees have a bluish or greenish-grey tint. This grey-green shade of the Western coffees is entirely deficient in those of Asia. The value of the berry in our wholesale markets is not, therefore, a fictitious quality, as some imagine, but is real, and depends first upon the texture and form of the berry or seed, secondly on the colour, and thirdly the flavour. The texture of the berry and form, termed "style" by the coffee brokers, is so well defined and palpable to the initiated, that at one view they

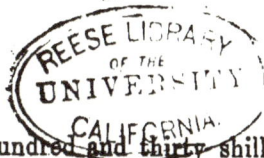

pronounce its value, from one hundred and thirty shillings per cwt. downwards, according to the two other qualities, colour and flavour.

The value of the coffees usually imported into this country stands in the following order: Mocha, fine Ceylon Plantation, Jamaica, Costa Rica, Java, Tellicherry, and St. Domingo.

Portugal produces coffee in several of her colonies. Ordinary description, yellowish berry, in St. Thomas; tolerably good in the Cape de Verdes; bad, yellow, in Timor; worse (but curious from the very small size of the berry), growing wild, in Mozambique; good in Angola; and of excellent quality in Madeira and Porto Santo, but the production is limited.

Much of the coffee which finds its way into England as genuine Mocha is, in reality, Malabar coffee, sent to ports of the Persian Gulf from Bourbon, and when thus naturalised, finding its way to Europe. But the coffee of India even now competes successfully with that of Arabia, in Bussorah, and other local markets, which the latter had for centuries commanded as its own.

It is curious to watch the progress of English enterprise. The energetic and ubiquitous Anglo-Saxons hold India, and here we see coffee from India triumphing over the famous berry of Arabia. The cultivation of tea also is rapidly spreading over 30,000 square miles of the Sub-Himalayan ranges; and who knows but that Indian teas may yet compete with those of the flowery land in the markets of Shanghae?

Already the Assam teas are held in high estimation by good judges of tea in this country, whilst they fetch a high price in India for local consumption.

The colour of the berry is by no means a decisive criterion of excellence of quality; in some parts the bluish berry is

C

esteemed most highly, in others the yellow. The West Indian coffees often change colour when kept a few years.

It is well known that the various sorts of coffee imported into Europe from the several parts where the plant is cultivated differ widely in quality and flavour. Levantine or Mocha still retains its old superiority in this respect, though the best sorts imported from Ceylon, Bourbon, Mauritius, and other Indian Islands are now generally considered to come very near it. This difference in quality and flavour of the various sorts of coffee is generally attributed to climatic and local causes and influences, which are necessarily beyond the power of remedy. This, however, is a great mistake. The more or less advanced state of maturity to which the berry is allowed to attain before picking, and, more particularly still, the degree of dryness, and the longer or shorter period of time for which it is kept before being sent into the market, exercise a most powerful influence upon the quality and flavour of the article. Berries gathered before they have attained maturity, though they may be perfect in colour, will always have a raw, herbaceous taste. If the drying berries are heaped too thickly or closely, they are apt to heat and to contract an unpleasantly bitter and harsh taste, and a disagreeable smell; this will frequently occur also where artificial heat is had recourse to to expedite the drying. Keeping tends to cure these serious defects in coffee. There are instances on record where coffee of a most disagreeable flavour and smell has been brought near perfection by being kept for some years in a dry loft; and though it may be going too far to assert, as has been done by some high authorities on the subject, that " the worst coffee produced in the West Indies will, in a course of years not exceeding ten or fourteen, be as good, parch and mix as well, and have as high a flavour as the best we have now from Turkey," still there can be no doubt that long keeping will most materially improve the quality of even the worst sorts. Unfortunately, the difference of price between inferior and superior coffee is not sufficiently great to cover so many years' interest on the capital invested. It is for the same reason that planters, though they are perfectly aware that trees growing on a light soil, and in dry and elevated spots, produce smaller berries of very superior flavour to those grown in rich, flat, and moist soils, yet prefer cultivating the latter, simply because the production is double that of the better sort. Those who wish to improve the quality of their coffees by keeping, must bear in mind that perfect dryness of the loft or warehouse, moderate warmth, and gentle ventilation are the indispensable conditions of success; a strong draught of air is more particularly to be guarded against, as it tends to bleach the berries. Great care must be taken, also, to keep all strong-flavoured wares, such as pepper, pimento, ginger, cod-fish, herrings, rum, &c., as far as possible from the coffee, which has a powerful attraction for these scents, and gets thoroughly impregnated with them, to the great deterioration, of course, of its quality. This remark applies more particularly to the shipping of coffee from Jamaica and the other West India Islands. Want of proper ventilation in the holds in which a cargo of coffee is stowed on board ship, is equally injurious to the quality of the article. Coffee which has suffered damage by sea-water, or has been spoiled by the close vicinity of strong-scented wares, may, to some extent, be reclaimed by "rouncing" or putting it in a tub, pouring boiling water over it, stirring for a few minutes, then pouring the water off, repeating the same operation a second, or even a third time, if necessary, and most carefully drying the washed berries.

SECTION V.

CHEMICAL ANALYSIS.

COFFEE has been analysed by several chemists, and though the results obtained differ in some slight degree, yet it seems pretty clear that the principal constituents to which its hygienic and medicinal properties are due are caffeine, a peculiar volatile oil generated in the roasting, and a kind of tannic acid.

The alkaloid caffeine, or theine, is found in one or two other plants besides tea and coffee. It occurs in the seeds of *Paullinia sorbilis*, a native of Brazil, and in the leaves of several species of holly, natives of South America, which furnish the Paraguay tea, or Yerba mate, so large an article of consumption in several of the South American republics. The leaves and young shoots, dried, parched, and pulverised, are used for a hot infusion. A kind of cake, called Guarana bread, is made from the seeds of the *Paullinia*, which is highly esteemed in Brazil and other countries when infused, like chocolate, for its nutritive and febrifuge properties, and is sold generally as a necessary for travellers, and as a cure for many diseases.

The nutritive and medicinal virtues of all these plants must certainly be attributed in a great degree to the presence of this chemical principle, and to the tannic acid which they also contain.

The use of coffee as a beverage has been examined in a chemical and physiological point of view by Professor Lehmann. The general results of his investigations are :

1. That a decoction of coffee exercises two principal actions upon the organism, which are very diverse in character, viz. increasing the activity of the vascular and nervous system, while at the same time it retards the metamorphosis of plastic constituents.

2. That the influence of coffee upon the vascular and nervous system, its reinvigorating action, and the production of a general sense of cheerfulness and animation, is attributable solely to the mutual modification of the specific action of the empyreumatic oil and the caffeine contained in it.

3. That the retardation of the assimilative process brought about by the use of coffee is owing chiefly to the empyreumatic oil, and is caused by caffeine only when taken in large quantities.

4. That increased action of the heart, trembling, headache, &c., are effects of the caffeine.

5. That the increased activity of the kidneys, relaxation of the bowels, and an increased vigour of mental faculties, passing into congestion, restlessness, and inability to sleep, are effects of the empyreumatic oil.

Professor Lehmann considers it, therefore, necessary to regard the action of coffee, and, in a less degree, that of tea, cocoa, alcohol, &c., upon the organism, as constituting an exception to the general law, that increased bodily and mental activity involves increased consumption of plastic material.

Caffeine, on careful analysis, has been found to contain in 100 parts, 49·80 of carbon, 5·08 of hydrogen, 28·83 of nitrogen, and 16·29 of oxygen. It is inodorous, but has a slightly bitter taste. The proportion in which this principle is found to be present in coffee varies between ¾ lb. and 1¾ lbs. in 100 lbs. of berries.

The peculiar essential oil which is generated in coffee in the process of roasting, by the action of heat upon some yet unascertained principle contained in the berry, is also very similar to the volatile oil in tea; but the quantity of it in coffee appears to be comparatively very small; for whilst 100 lbs. of tea-leaves contain 1 lb. of volatile oil, it takes 500 cwts. of roasted coffee to give a similar quantity; and yet it is upon the presence of this oil that the flavour and value of the several varieties of coffee mainly depend.

The tannic acid is, by some chemists, also said to be generated only in the process of roasting; others maintain that it is present in the raw bean.

The chemical properties of the coffee-berry are altered by roasting, and it loses about twenty per cent. of weight, but increases in bulk one-third or one-half. Its peculiar aroma, and some of its other properties, are due to a small quantity of essential oil, only one five-thousandth part of its weight, which would be worth about 100l. an ounce in a separate state. Coffee is less rich in theine than tea, but contains more sugar and a good deal of cheese (casein).

Schrader has analysed raw and roasted coffee, with the following result:

	raw.	roasted.
Peculiar coffee principle	17·58	12·50
Gum and mucilage	3·64	10·42
Extractive	0·62	4·80
Resin	0·41	
Fatty oil	0·52	2·08
Solid residue	66·66	68·75
Loss	10·57	1·45
	100·00	

" The examination of coffee," observes Dr. F. Knapp, " has led to interesting results, although they are still defective in pointing out the quantitative composition of the berry."

The following is the composition of the ash according to Levi:

Potash	50·94
Soda	14·76
Lime	4·33
Magnesia	10·90
Oxide of iron . . .	0·66
Phosphoric acid . . .	13·59
Sulphuric acid	trace
Chlorine	1·22
Silicic acid	3·58
	99·98

According to the analysis of Payen, the unroasted coffee-berry has the following composition:

Moisture	12·0
Glucose and dextrine	15·5
Nitrogenous matters	13·0
Chlorogenate of caffeine, &c. . .	3·5 to 5·8
Fatty substances . . .	10 to 13·0
Cellulose and woody fibre	34·0
Mineral substances in ash	0 7
Essential oil	·003
	100·0

Or to define the per-centage more closely, we may put it thus:

Water	12·000
Caffeine, or theine	1·750
Casein	13·000
Aromatic oil	0·002
Sugar	6·500
Gum	9·000
Fat	12·000
Potash, with a peculiar acid . . .	4·000
Woody fibre	35·048
Mineral matter	6·700
	100·000

In another form this shows us:

Water	12.00
Flesh-formers	14·75
Heat-givers	66·25
Mineral matter	7·00
	100·00

As gluten is only very sparingly soluble in boiling water, in the usual way of making coffee the flesh-formers are thrown away with the dregs; the addition of a little soda to the water partly prevents this waste.

The various components in one pound of coffee will be—

	oz.	grs.
Water	1	407
Caffeine, or theine		122
Casein, or cheese	2	35
Aromatic oil		1½
Gum	1	192
Sugar	1	17
Fat	1	402
Potash		280
Woody fibre	5	262
Mineral matter	1	31

The part roasted is the albumen, which is of a hard, horny consistence; and Lindley remarks that it is probable that the seeds of other plants of this or the stellate order, whose albumen is of the same texture, would serve as a substitute. This would not be the case with those with fleshy albumen.

Coffee loses in weight by roasting, but gains in bulk in proportion to the heat applied.

Payen found the following amount of nitrogen in 100 parts dried:

	nitrogen.		ash.
Martinique	2·46		5·00
Bourbon	2·54		4·66
Mocha	2·49		7·84

The coffee from Martinique lost 11·58 per cent. of its weight by drying. This description of coffee also afforded the following results :

	Unroasted.	Slightly reddened.	Chesnut brown.	Brown.
Loss in roasting	—	15 per cent.	20 per cent.	25 per cent.
Increase in bulk	—	1·3 times	1·53 times	—
Extract	40 per cent.	37 per cent.	37·1 per ct.	39·25 per ct.
Insoluble residue.........	48·5 „	—	—	—

Coffee, as ordinarily prepared for beverage, contains only two-sevenths of the nitrogenous or nutritive matter of the fresh bean, but two-thirds of the roasted, and the mineral ingredients are all present.

M. Lebreton (" Agriculteur praticien") has estimated the loss of weight of coffee in roasting at 18 to 20 per cent. in Porto Rico, Rio, and Martinique coffee; and at 16 to 18 per cent. in Malabar, Bourbon, Ceylon, and Guadaloupe coffees; while in Mocha coffee it amounts to only 14 or 10. The loss of weight depends upon the time of roasting and the degree of heat. Damp or damaged coffee loses more than dry sound coffee. He considers that these substances have the capability of rendering the individual insensible of a certain deficiency of food, in virtue of their retardation of the assimilative process. He thinks it probable, likewise, that these substances have a direct nutritive value, especially coffee as drank by the Turks and Arabs with the grounds.

Professor Lehmann considers that the singular preference for one or other of these beverages by particular nations, as well as the Eastern custom of drinking coffee with the grounds, are not accidental, but have some deeper reason. This reason, he thinks, is to be found in the different effects of the coffee, tea, &c., and the various requirements of the nations by whom they are used, and instances the use of tea by the English, and of coffee by the Germans and French, as in accordance with this view. The diet of the former affords a larger supply of plastic material than that of the latter people ; and while, consequently, the retardation of the assimilative process is an important influence for the German, the proportionately greater nervous stimulus caused by tea is more desirable for the former. The use of coffee with its grounds has its analogue in the use of tea mixed with meal, milk, and butter among the Mongols, and other inhabitants of the Central Asiatic steppes.

M. Payen, from elaborate experiments, shows that coffee slightly roasted is that which contains the maximum of aroma, weight, and nutrition. He declares coffee to be very nourishing, as it contains a large quantity of nitrogen, three times as much nutriment as tea, and more than twice the nourishment of soup. Chicory contains only half the nutriment of coffee.

SECTION VI.

COFFEE-LEAF TEA, &c.

ATTENTION was some time ago drawn to the subject of coffee-leaf tea, which is used in Sumatra and other parts of the East, and a good deal of discussion ensued upon the matter, after the leaves were shown for the purpose at the International Exhibition of 1851. An infusion of roasted coffee-leaves is pronounced by those who have had an opportunity of tasting it, as superior to Bohea, and by some enthusiastic admirers is said to rival the flavour of the most delicate Pekoe. That an infusion of roasted coffee-leaves should imitate the flavour of tea is not to be wondered at, as the leaves of both shrubs contain in the main the same leading principles, more particularly theine or caffeine. There is no doubt that coffee-leaf tea would command a sale in England, but the question is how much could be collected to make it profitable, and it involves the necessity of apparatus and skilled labour for parching the leaves.

Coffee-leaves are not quite so thick as those of Vallambrosa, and a Malabar coolie would not in one day collect enough to pay the expense of picking, drying, packing, cartage, warehouse rent, freight, and other charges.

Moreover, no planter of any experience would think of stripping his trees of their breathing organs, and the quantity that might be collected from the suckers and prunings, &c., would never give more than a few bales, even on large plantations. Even were the fallen leaves supposed to be available, their removal would be as detrimental as the practice of raking away withered leaves in plantations, or the application of the sugar-cane trash to the purposes of fuel.

The husks, pulp, and parchment in South America, the West Indies, Ceylon, and the other Indian islands, are regarded as mere waste, and thrown away. In Arabia and some parts of the East, however, this refuse is utilised, as the "miserables," or husks of the cocoa-seed (*Theobroma cacao*), are in Ireland and the Continent. With it is prepared the famous kisher, or "Café à la Sultane," a light-coloured, bright infusion, which has all the agreeable flavour of coffee, with little of its strength and none of its bitterness; this is partaken of by the humbler classes in incredible quantities.

When quite dry and ripe these husks are bruised, and roasted in an earthen vessel over a charcoal fire, not as coffee usually is, but only until it assumes a light-brown colour. While hot it is thrown into a pot of boiling water, with a small proportion of the pellicle or parchment skin; all is boiled together for a few minutes, and then served hot and strong, but without sugar. Sometimes a drop of essence of amber is put into each cup; or cloves, aniseed, or cardamoms are boiled with it.

In Brazil, from the sweet pulp which envelopes the berry an excellent spirit has been made.

SECTION VII.

ADULTERANTS.

GREAT as the consumption of coffee is in Europe and the Americas, it has become so necessary in every household, that the demand continues to increase, and very full prices are maintained. The largely extended plantations in Brazil, in Ceylon, in India, in Java, and in other suitable localities, profitably opened up every day, altogether fail to keep down prices, and will long continue to offer the strong inducement of large profits!

Taking advantage of this great public want, the unscrupulous and fraudulent trader foists upon the easily-duped masses of consumers vast quantities of unwholesome and pernicious stuff, mixed in certain proportions with coffee; and, although the law has interposed in the case of chicory, forbidding, under penalties, its being sold mixed with coffee, unless especially so.labelled and declared, yet who can tell the thousand and one mixtures that are still made and sold with it?

It may be said that all this can be prevented by purchasing only unground coffee, in a roasted state, ready for grinding; but there are those (to be reckoned by hundreds of thousands of families) who, having no means of grinding this roasted coffee, are compelled to buy that which is already ground, or go without altogether.

What abominations do these (in too many cases) not drink, then, under the much abused name of *coffee?*

These reflections are forced upon my mind every time I enter a coffee-house, and am called to put faith in the purity of the cup of coffee set before me.

Visitors to the establishments of coffee-grinders speak of

bags of the husks of peas, and cobs of Indian corn, wheat, sea-biscuits, and other articles, harmless enough in themselves, and in their right places, where, indeed, they may be useful; but a coffee-mill is the wrong place for these otherwise objectionable articles of consumption. It will not be denied that the husks of peas or the cobs of maize are appropriately placed in the trough from which our pigs feed; they will even add to the delicacy and whiteness of the pork those very useful animals are intended to yield us, but they are essentially out of place in our coffee-cup. These, and various other articles, go to make what in the trade is termed "Boston," and this preparation the visitor, who has been admitted to the arcana of the grinder, will find mixed in a large bin ready for use. One who has been thus favoured, after describing the mill used for grinding the coffee, speaks of another machine, called the "mixer." The "mixer" is a wooden cylinder revolving on a spindle at an angle of 45°, and having internal arrangements to mix the coffee with such proportion of adulterating material as the ingenuity or impudence of the grinder may suggest. Here, like Macbeth's witches over their cauldron, presides the genius of the place, who casts in the ingredients which constitute the villanous compound, which, done up in tins, is finally ornamented with labels such as this:

FINEST PLANTATION COFFEE,
Roasted and ground on the most approved principle by
STEAM MACHINERY.
This coffee is with confidence highly recommended to the public, as the fine aroma, so much appreciated, is entirely preserved by its being
PACKED FRESH FROM THE MILL.

We need scarcely say that the amount of nourishment contained in this abominable *olio* is infinitesimally small.

The *Scientific American* of New York, of a recent date, records the following fact: "The editor of the *Baltimore American* lately visited the commissary department of one of the large military hospitals, and noticed several barrels of dried coffee grounds, the purpose whereof excited his curiosity. The polite commissary informed him that they received twelve dollars a barrel for the grounds. 'But what is it purchased for?' he asked. 'Well,' said the commissary, hesitatingly, 'it is re-aromatised by the transforming hand of modern chemistry, and put up in pound papers, which are decorated with attractive labels and high-sounding names.'"

About ten years ago, when the question of coffee adulteration was much agitated, I published a little treatise, entitled, " Coffee, as it is and as it ought to be," in which, among other particulars, I pointed out the various sophistications practised. A compositor who was engaged in printing the work furnished me with a remarkable statement in confirmation of reports which I had previously heard.

He stated that in various parts of the metropolis, but more especially in the east, are to be found liver bakers. These men take the livers of oxen and horses, bake them, and grind them into a powder, which they sell to the low-priced coffee-shop keepers, at from 4d. to 6d. per pound; horse-liver coffee bearing the highest price. This adulterant may be known by allowing the coffee to stand until cold, when a thick pellicle or skin will be found on the top. It goes farther than coffee, and is generally mixed with chicory and other vegetable imitations of coffee.

According to the investigations of Dr. Hassall ("Food and its Adulterations"), the several adulterations of coffee may be distinguished by the following characters :

Chicory, by the size, form, and ready separation of the component cells of the root, as well as by the presence of an abundance of spiral vessels of the dotted form.

Roasted corn, by the size, form, and other characters of the starch granules, of which the grains are principally composed. Beans, also, by the form, &c., of the constituent granules of starch. Potato, by the large size, rounded form, and ready separation of the cells of the cellulose, as well as by the fibrous markings on their surfaces.

Plate 3 shows a fragment of roasted coffee as seen under the microscope, magnified 140 diameters, and the structures in a sample of coffee adulterated with chicory.

Dr. Normandy, in his evidence before the Parliamentary Committee on the Adulteration of Food, &c., stated that he had met with roasted corn in coffee to an extent of from 25 to 30 per cent; it is recognised by the size and character of the starch granules; consists of barley and rye, and is generally very easy to detect; it floats up. If coffee has been adulterated with roasted grain, when you pour boiling water upon it you will see rising against the sides of the cup portions of the ground grain, which you sometimes can separate in considerable quantities, by capillary attraction. If you pour such coffee from the coffee-pot, some of those grains will fall with the liquid in the cup, and they will climb up, as it were, the sides of the cup a quarter of an inch, or something of that kind, all round, and you can collect them very easily.

Dr. R. D. Thompson, F.R.S., another witness, stated that a large cargo of lupins from Egypt having been imported, which could not be made any use of in consequence of their bitter taste, he was asked to give a certificate in favour of their being equal to coffee, but declined, and recommended that, after steeping, to remove the bitter principle, they should be sold for cattle. He also added that he had seen an ingenious apparatus for making artificial coffee-berries from chicory and other substances; it was something like a bullet-mould, and patented by Messrs. Duckworth of Liverpool.

The chicory itself sold is not always pure. The Board of Inland Revenue have found, on chemical examination in different samples, beans, rye, oats (roasted and ground), caramel or burnt sugar, oxide of iron and orange berries.

The principle contained in coffee, remarks Dr. Letheby, may be considered as essential to life, inasmuch as all nations

Plate 3.

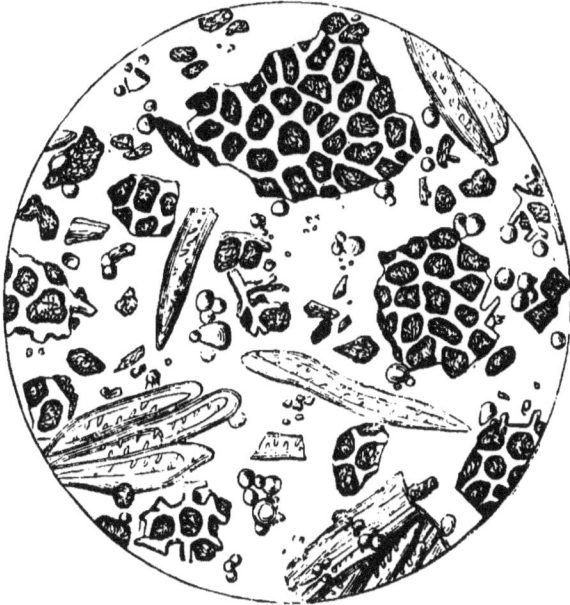

Genuine Ground Coffee.

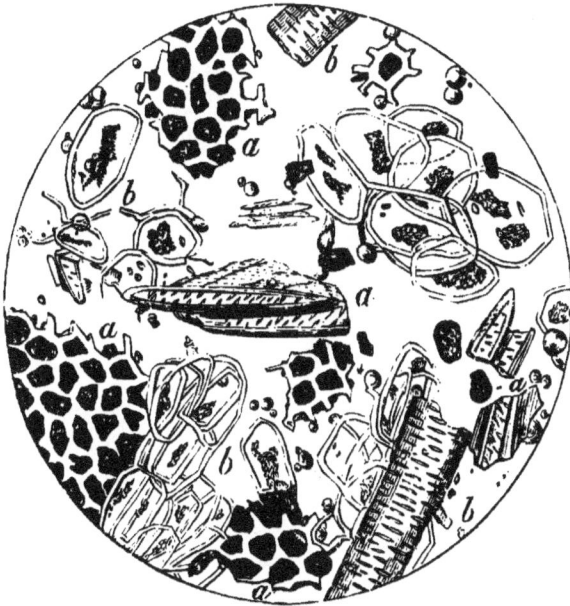

Coffee Adulterated with Chicory

use it; it prevents the wear and tear of the body. Now there is nothing in chicory which can do that. Chemically there is no difference between beetroot chicory and what is called real chicory; microscopically you can discover the difference. If genuine coffee is sprinkled upon the surface of a tumbler of water, it remains a considerable time floating, and when it sinks it only slightly colours the water; chicory, on the other hand, sinks quickly, and colours the water very deeply. Ground coffee is enveloped in an oily substance, which prevents the water absorbing ; chicory has no such protection, and sinks immediately.

Mr. Phillips, the chemist to the Inland Board of Revenue, states that the average per-centage of chicory to coffee, when sold mixed, was found to range from $20\frac{1}{2}$ to $16\frac{3}{4}$ per cent.

At a recent meeting of the British Association of Science, Mr. Horsley called attention to the use of bi-chromate of potash, in analysing adulterated samples of coffee. With diluted solutions of pure coffee, this salt produces an intense deep porter-brown coloration, whilst upon decoctions of chicory no effect is produced. He advised the following procedure : Take equal parts of chicory and coffee, and decoct them in different quantities of water. Filter, bottle, and label the liquids. Take a teaspoonful of the chicory, and dilute till it is of a brown sherry colour ; boil it in a porcelain dish, with a fragment of crystallised bi-chrome. The colour will be scarcely deepened. If a similarly diluted solution of coffee is thus treated, a deep-brown tinge is obtained. By operating with mixed liquids a scale of colours may be obtained indicating the properties of the two substances. If a few grains of the sulphate of copper be added, both decoctions yield a precipitate; that from chicory being a clay yellow, and that from coffee a sepia brown. · Mixed decoctions yield intermediate tints.

D

SECTION VIII.

CULTURE IN THE WEST INDIES AND AMERICA.

In 1720, a small coffee-plant, raised in the garden of the King at Paris, was transported to the Antilles by Captain Declieux, who, during a long passage, shared each day his small allowance of water with the young coffee plant. From this tree have sprung all those since cultivated in Martinique, Guadaloupe, Cayenne, St. Domingo, and the other islands. The fall of St. Domingo, in 1789, which formerly furnished 80,000,000 lbs., the disfavour the culture has fallen into in Martinique and Guadaloupe islands, which used to supply 16,000,000 or 17,000,000 lbs., together with the greater attention given to sugar cultivation in the West Indies, have transferred the production of coffee chiefly to Brazil, Ceylon, and Java.

In 1801, 526,000 cwts. of coffee were imported into Great Britain from the West Indies, and the average annual imports in the six years ending 1806 was 364,000 cwts. The decrease of production is shown by the following figures :

	1829.	1850.	1860.
	lbs.	lbs.	lbs.
Jamaica	18,690,654	4,196,210	6,145,362
British Guiana	7,163,016	18,472	—
Trinidad	73,667	96,876	—
Dominica	942,114	792	10,000
St. Lucia	303,499	39	—

JAMAICA.—The coffee-plant is said to have been first introduced into Jamaica by Sir Nicholas Lawes, in 1728, when it was cultivated on an estate called Temple Hall, on the plains of Liguanea, not far from Kingston. In 1752, 6,000,000 lbs. of coffee were exported, and in the three years ending 1807, the average annual shipment was 28,500,000 lbs. It was not until the island trade was injured by the dismemberment of Honduras that the heavy duties on sugar, and the competition of the French colonies, induced the industrious planters to turn their attention to coffee. They then petitioned for a protecting bounty, and for a considerable time this was the only British colony where its cultivation was much attended to. In 1791 there were 607 coffee-plantations in the island, with 21,000 negroes employed on them. In 1844, there were 671 plantations, and it was calculated that 20,000,000l. sterling was invested in them. Some of the finest and most productive plantations were at an elevation of 4700 feet, in the vicinity of the Blue Mountain Peak. Whilst in 1809, 83,250,000 lbs. of coffee were shipped from Jamaica, the average export for ten years past has not exceeded 6,000,000 lbs.

In the West Indies and Central America a nursery of young plants is usually raised from seed. At the age of six months these are transplanted into the field in squares of 8 to 9 feet apart. Beyond keeping the field clear of weeds, giving the plants a slight moulding, and freeing them of suckers or sprouts, they do not require any other attention for the first three years. At the expiry of this time they commence to bear fruit, and the coffee is cured either by drying the entire fruit in the sun, or by passing the fruit through what is called a pulping mill, which separates the outer pulp or covering from the beans. The beans are then subjected to a washing in a tank full of water, in order to remove the glutinous substance which adheres to them; this causes them to cure or dry more rapidly. They are then left in what is called the parchment husk, and exposed to the action of the sun on platforms laid with tiles, until they are perfectly dried, and from this they are, during crop time, spread on the wooden floors of a building called a logie, to the depth, in a heavy crop, of 12 to 18 inches. Whilst thus spread, great attention is necessary to prevent the coffee getting heated and its colour destroyed, by constantly turning it up, and by the use of some dry lime or ashes sprinkled over it. After this the beans are subjected to the operation of what is called a stamping mill, which separates the parchment husk from the beans. This stamping mill is merely two large solid wheels fixed on each end of a beam on an axle, and worked by mules moving round in a circle—the wooden wheels working in a circular trough filled with the coffee beans. After this the whole are submitted to the action of a winnowing machine, which separates the chaff from the beans, and subsequently the beans are passed through copper sieves to separate the perfect from the broken coffee, and finally handpicked and put into bags or casks for shipment.

BRITISH GUIANA.—Coffee was for a length of time almost the only staple of Berbice and Demerara, but the cultivation of the sugar-cane has been substituted for it. The quantity of coffee, the produce of British Guiana, exported in 1830 was, 9,472,756 lbs. ; in 1840, 3,357,300 lbs. ; in 1849, 100,550 lbs.; and in 1850, 30,000 lbs.; since which it has almost ceased to be exported, scarcely sufficient being produced to supply the demand in the colony.

At one time there were about two hundred coffee plantations in the small island of DOMINICA, and four to five million of pounds of coffee were exported annually to Great Britain.

PORTO RICO.—In proportion to its extent, this island is twice as productive as Cuba, and the quality of its coffee and other produce is of the highest class. There were, in 1862, fifty-three coffee plantations on the island, producing about 100,000 cwt. of coffee. The exports of coffee were, in lbs.

1855	. . . .	11,506,283
1856	. . . .	9,935,000
1857	. . . .	8,245,000
1858	. . . .	9,814,000
1859	. . . .	13,457,000

The exports of coffee from the island of CUBA, which were in 1840 upwards of 2,000,000 arrobas, were in 1858 but 21,000; and in 1859, 5000 or 6000.

BRAZIL.—The coffee-plant has been known in Brazil for many years; it is but about fifty years, however, since the first regular plantation was made by Mr. Moke, a Belgian, who brought the cultivation of coffee to great perfection. His plantation is still in the neighbourhood of the capital, and is carried on by his son with much success. It is astonishing to what an extent coffee has been cultivated since Mr. Moke first made his plantation. Two millions of bags of 160 lbs. each are annually exported from Rio de Janeiro, taking the average of the last seven years; 1862 was, however, short of this about half a million bags. At Barahyba do Sul, which is within a few miles of Rio, there are plantations employing six and seven hundred slaves.

The best plantations are those owned and conducted by foreigners—chiefly English, French, and Belgian—they have an air of neatness and comfort about them of which those owned by Brazilians and Portuguese are totally destitute. The foreigners use improved machinery also in preparing the berry for market, which the Brazilians, with some exceptions, do not. The coffee-berry contains two seeds, covered with a gummy, mucilaginous substance, and enclosed in a skin which is thick, sweet, and dark and red when ripe. The foreigners take off this skin by means of machinery, and the beans are washed until they are divested of the mucilage which covers them. They are then dried and put into bags ready for market.

The Brazilians dry the beans with the skin on. In the process of drying, the skin first becomes dark, and finally black, and when crisp, is rubbed off the bean, which is then washed. In this process, however, there is great danger of fermentation. The skin contains a vast amount of saccharine matter, and successful attempts have been made to extract from it sugar and spirit; but either through poor machinery, or other mismanagement, it was found to be unprofitable, and the experiment was abandoned. The skin is exceedingly sweet, almost as much so to the taste as the sugar-cane.

The coffee-plant can be propagated from the seed, but the most prevalent method in Brazil is by young plants, which may be had by the thousand on old plantations. The young shrub is taken up in August—generally when it is about two years old—and planted in good soil. The fourth year it produces coffee, and the fifth year it commences to bear regular crops, the yield being from a pound and a half to three pounds per tree. Trees have been known to last for many years on good rich soil, and some on Mr. Moke's plantation are still bearing which were planted forty years ago; on hill-sides, however, where the soil is light, the plant decays in the course of eight or ten years. The picking season commences in July, and in the low lands generally concludes by the end of August; among the hills, however, where there are frequent showers, and where there is much shade, the season does not close until some time in September.

The four coffee-growing provinces of Brazil, properly speaking, are Rio de Janeiro, San Paulo, Minas, and Bahia. In the two former the agriculturists have for some time past directed their attention almost exclusively to this article. In Bahia its cultivation is steadily, although slowly, increasing. Coffee is grown in the provinces of Maranham, Parana, and St. Catherine's, but only to a very small extent, the amount

produced being insufficient to meet the demand. The culti-vation of coffee is rapidly increasing in St. Paulo, and is promoted by the railroad extending from Santos to Jun-diahy. The United States usually takes one-half of the crop of Brazil coffee.

The following figures give the average quantity of coffee exported from the empire of Brazil during the undermentioned years, comprising a period previous and subsequent to the abolition of the slave trade in 1851, and showing a steady increase in the export since 1840:

	Annual average. arrobas.
1840-43	5,507,367
1843-46	6,519,380
1846-49	9,301,967
1849-52	8,542,965
1852-55	10,549,847
1855-58	11,465,719
1858-61	10,501,665

The yearly shipments from Rio in the last four years have been as follows:

	bags.
1859	2,064,837
1860	2,150,188
1861	2,085,974
1862	1,477,904

besides 200,000 to 300,000 bags annually from Santos. The bags contain rather more than six arrobas, or one cwt. and a half.

COSTA RICA.—The quality of this coffee is recognised as excellent. In order to place it in a proper degree of esti-mation in foreign markets, some proprietors of plantations have not omitted getting it chemically analysed in Europe;

the result of which examination has classed it in the third degree, among those kinds generally esteemed as the best.

The following figures show the production of coffee in Costa Rica; the bulk of the crop is sent to Great Britain:

	cwts.
1845	70,000
1855	70,709
1856	83,000
1857	95,000

VENEZUELA.—Within the last thirty years Venezuela has made great progress in coffee culture. The exports, which were not more than 13,000,000 lbs. in 1833, had risen to 15,000,000 lbs. in 1850; but since then the culture has been much interfered with by civil war. The exports from La Guayra were, in 1855, 17,375,000 lbs.; in 1856, 12,357,000 lbs.; and in 1857, 16,031,000 lbs. Washed coffee sells there at 8s. per cwt. higher than the unwashed.

In Venezuela, the plan followed in drying the coffee is this. The berries are spread out upon hurdles in the sun, where they undergo the vinous fermentation from fourteen to twenty days, and then dry. The beans are freed from the pulpy husk by a mill in two operations, and from the parchment, &c., by winnowing. Although a single tree may bear as much as 10 to 20 lbs., the average in Venezuela is under 2 lbs. An acre planted with 2560 trees yields there an annual crop of about 1100 lbs. of dry beans.

In ECUADOR coffee has of late greatly engrossed the attention of planters, from its superior quality and the increased demand; it commands a higher price than any other product of the country in proportion. As the people are everywhere dedicating themselves to its culture, in a few years it is probable that this country will export a considerable quantity.

In 1855, 776 cwt. only were shipped; in 1861, although the harvest was exceedingly scanty, the exports were 1480 cwt.

GUATEMALA.—Some twenty years ago considerable plantations of coffee were made in different parts of the republic of Guatemala, and had the undertaking been followed up with proper and steady constancy and attention, it might, perhaps, by this time, have become an important article of export. Unfortunately its culture was abandoned, owing to the insurrections of the Indian population. Some attempts have been made again to introduce its culture by making fresh plantations, but hitherto not to any extent worthy of mention. The exports of coffee in 1860 were about 63 tons. In the department of Vera Paz, there are said to be now about half a million of coffee plants in bearing, and nurseries of about a million more plants will soon be ready for transplanting. The greater part of the plantations are situated in the neighbourhood of Coban, where the vicinity of Teleman, on the river Polochec, which disembogues into the gulf of Dulce, affords a convenient channel for shipping the produce.

SECTION IX.

CULTURE IN ARABIA.

THE culture 'of coffee is principally carried on in Yemen, towards the districts of Aden and Mocha. The nearest coffee plantations are about eighty miles from Aden.

The coffee-growing country in the Yemen is 300 miles to the south of Jeddah, being the districts about Tohira, Uodeida, and Tanna. Of this, the Turks have but little, their authority only extending over the narrow slip called the Tehana, in which is little coffee. They had till lately the export dues of all the coffee grown in Western Arabia, but they have lost a great part of them, a large quantity of coffee being sent to Aden for exportation. Mocha, the ancient port and capital, has completely fallen, and is in ruins. Its place is taken by Uodeida, the seat of government of the Yemen, and a Pashalic under Jeddah. In 1860, coffee to the value of 14,268l. was imported into Aden by sea, of 55,710l. was brought by land, and 45,344l. was exported. An enormous quantity is consumed in Arabia. Very little genuine Yemen coffee is procurable in Europe. It is difficult to obtain even in Jeddah. It is first mixed in the Yemen with inferior Abyssinian coffee, then mixed at Jeddah with damaged coffee, and probably in Egypt it is again mixed. Alexandria and Cairo are notorious for bad coffee.

It has been understood for several years that much of the coffee which finds its way into England as genuine Mocha is, in reality, Malabar coffee, sent to ports on the Persian

Gulf from Bombay, and when thus naturalised, finding its
way to Europe. But the coffee of India is now competing
successfully with that of Arabia in markets which the latter
had for centuries commanded as its own. In a few years the
Indian article will entirely surpass the Arabian. It is the
old story of enterprise originated and directed by Europeans
driving competition out of the market. Seeing that the
Arabs are in the habit of baking the cow-dung and cakes for
food on the same wall, they cannot be suspected of any
violent antipathy to dirt *per se*. But, passionate lovers as
they are of a good cup of coffee, they can easily understand
that an equal quantity of well-cleaned Malabar coffee is
cheaper than that brought in the buggalows of the Nijd
Arabs, and which is described as full of extraneous matter,
such as pieces of the coffee-husk, &c. But the Indian coffee
is actually sold cheaper, quantity for quantity. The consumption of Yemen coffee is now entirely confined to the
wealthier chiefs; the finest kind, and that most sought after,
being the small, of a light-green colour. Latterly, the
Yemen coffee has been all sent to Bagdad, a compliment
which the ancient city of the Sultans cannot fail to appre-
ciate at its true value. It is too dear and too dirty for the
poor Arabs of Bussorah, while the flavour of the Malabar
kind is found closely to approximate to that of Yemen.

The coffee is generally grown half way up the slopes of the
hills, but some is cultivated on lower ground, surrounded by
large trees for shade. The harvest is gathered at three
periods of the year, the principal being in May. Cloths are
spread under the trees, which are shaken, that the ripe fruit
may drop. The berries are then collected and exposed to the
sun on mats to dry. A heavy roller is afterwards passed over
the berries, to break the envelopes, and the husk is win-
nowed away with a fan. The berries are further dried before
being stored.

Arabian coffee was first cultivated by the Dutch, and some of their plants sent by the French to the West Indies, where it is now successfully cultivated. So that one has a difficulty in deciding where it is indigenous, as it is a much more important article of agriculture in the West Indies, Java, Ceylon, and Southern India, than in its native countries.

In Syria the coffee-plant is of natural growth; but as the European writers who were engaged in the Crusades do not mention it, it could not have been much used during the twelfth and thirteenth centuries. Bruce affirms that the qualities of it were well known in Africa, and that the Gallae, a wandering tribe, which was obliged to traverse the deserts, carried no other provision than balls compounded of coffee and butter, one of which would keep them in health and spirits through a day's journey better than any other kind of food. In the Royal Library at Paris is an Arabian manuscript, containing a voluminous history of coffee, in which it is said that Gemaleddin-Ahou-Abdallah, Mufti of Aden, first introduced its use among the Turks, upon his return from Persia, where he had experienced the beneficial effects of it as a common beverage. The effendi, the kadi, and all the inferior officers of the government, followed the example of this chief of the law. The use of coffee descended through the harem to the house of every merchant, and the town of Aden set the example to the rest of Arabia.

SECTION X.

CULTIVATION IN CEYLON.

In 1824, Sir Edward Barnes and Sir George Bird commenced coffee-planting in Ceylon on a large scale. Others followed gradually, but the real rush for land dates from 1833. There are now 250,000 acres owned by coffee-planters, of which 100,000 are cleared and cultivated.

In Ceylon coffee succeeds best at an elevation from 1200 to 4800 feet; the quantity, generally speaking, lessening, but the quality improving with the elevation. By the natives a little of inferior description is grown in the low country. In 1855 there were about 150 estates belonging to Europeans, comprising 30,000 acres of cultivated ground, to which considerable additions have since been made. 400,000 cwt. was exported in 1853, and 783,393 cwt. in 1863.

The railroad forming from Colombo to Kandy, and the opening up of new roads, have rendered more land accessible, and the large clearances made in the forest at the high elevations will probably so improve the climate of those localities that a still further addition to the available land will be obtained. It is, therefore, not too much to expect the export, which in the last five years has on the average exceeded 630,000 cwt. per annum, will in less than a quarter of a century be more than doubled. In the fourteen years ending with 1862, Ceylon has sent into the markets of the world the following quantity of coffee:

	cwts.		£
Plantation	4,625,995	of the value of	11,310,518
Native growth	1,945,623	„	3,492,290
	6,571,618		14,802,808

Elevation must ever be an important consideration; though we have no doubt that by the use of manures coffee might be made to do well and bear to a limited extent at the level of the sea in Ceylon. We believe this to be, however, so completely artificial, that it will never again be tried whilst land is available from 1500 feet and upwards of elevation, where the plant grows vigorously without more than ordinary care being bestowed. The general effect of elevation may be described very shortly; the lower ranges, with fair soil, produce the heaviest crops and the soonest after planting, whilst the plantations on higher elevations produce smaller crops, but a finer quality of produce, and take a longer time to come into full bearing. There are many circumstances which go to produce climate, besides differences of elevation. On a plantation formed in a large district of forest yet uncleared, the climate will be colder and more moist than when the formation of other estates has cleared away the forest around it. The proximity of a high mountain peak, or being situated on the shoulder of a mountain which towers to a great elevation above the level of the plantation, will also produce more cold and wet than if the garden were opened on a lower range and to the full height of such a position. Plate 4, for instance, shows such a locality in the view of Konda-galla estate, near Neura-ellia, Ceylon. At low elevations long continued dry weather is more frequent than on the higher ranges, inasmuch as the clouds are frequently broken by the distant hills before they reach; and when there is no rain about, estates on the higher ranges are less parched, and atmospherically enjoy a moisture of which the lower hills are destitute. Still these various conditions are only advantageous or hurtful under varying circumstances; high elevation, cold and wet, and low elevation, heat, and drought, are alike unfavourable as prevailing characteristics; it is their due and seasonable admixture

Plate 4.—Konda-gala Estate, near Neura-ellia, Ceylon.

which is found most favourable to the profitable production of the coffee-plant. Perhaps the least favourable localities are those high positions where the natural vegetation is of the alpine character; in such positions the plant will only struggle for an existence; it therefore follows that the land selected should be well under these extreme elevations.

Soil of all kinds has had its advocates, and in turn been condemned by all. Depth and freeness are perhaps its most favourable states; dark black mould is always good; but wherever there is sufficiency of depth found beneath a virgin forest, at the proper elevation and climate, coffee has come on successfully. The land which has by general opinion been condemned as unfit for continued production is that covered with small jungle after the original forest has been cut down, and the land made to produce a rich crop for the natives. Perhaps it may be the exposure of the soil to the sun, and exhaustion by the large weeds which it produces, which are so injurious to it. The modern system of manuring is rapidly equalising the value of all soils, though deep lands on a limestone bottom, or strewed with granite boulders, are always considered highly favourable for coffee cultivation. The proximity of land to roads is a point of great importance in its selection, both as directly and indirectly affecting the outlay in forming a plantation, and probably for a long time determining the large item of expense under the head of transport of produce.

Felling and clearing, as the name implies, is cutting down the jungle and burning it off, so as to leave the land clear for planting. It is conducted as follows. Beginning from the lower part of the land and working upwards, the undergrowth, or small jungle, is cut down with catties or bill-hooks, leaving the forest trees free and open for the labours of the axemen. Likewise beginning from the lowest part of the field, the labourers, generally village Singhalese, who are extremely expert at this work, cut the trees nearly through by notches at the lower and upper sides, gradually retiring up the hill, until a tree of larger dimensions is cut, and being sent down, crashes all the others beneath it. When this is cleverly managed, several acres will often be opened to the daylight at one time; much, however, depends upon the steepness of the ascent and the heaviness of the forest. On the highest elevations the trees are smallest, and come down lightest, and on the lower elevations it may frequently happen that for acres and acres the trees are of that immense size that every log has required four men to cut it from the stump. Not unfrequently some of these trees have such projecting roots that the axemen have to erect stages around them to reach the ordinary trunk, which will each give employment for six hours to four men to cut through. After the trees have been felled, the lopping has to be attended to; this is to cut off all the tops and branches, and in some cases to cut the trunks across, so that the mass cut down may lay compactly and dry, as upon this depends much of the success of the burning, and, therefore, the economy of the operation. Small or light forest is often the most expensive to lop, from the lifeless fall of the trees, and the comparatively greater quantity of head and branch; whereas tall and heavy forest trees fly as it were under the axe from the stump, and in falling break themselves and all beneath them; in this manner some heavy forests cost less to clear than a lighter growth. The clearing having been left from six weeks to about two months, according to the weather, is fired, which is done generally by setting fire to it chiefly at *the lower ends* in several places; by this means the fire is soon connected, and burning in the dryest or first cut down portions, unites in a sweeping flame, rushes up the hill, destroying all before it; such is the power of the flame from below, that when the burning is successful the part last cut, probably many acres, and yet lying green upon the ground, is consumed with the rest. After the flame has passed away, nothing should be seen but the smoking black logs of the large trees, which, wherever they cross each other and do not lay upon the ground, require to be cut across and brought down. If the operation has been successfully managed, the land is now in readiness for the next operation—*Lining*, or staking out the positions of the intended plants. This is performed in many ways, but it is essential to have it done well, that the plants should make perfect lines across the ground to be planted, in order to maintain regularity in the plantation work, and give the best possible appearance to the field ; therefore no expense necessary to ensure its correct and workmanlike performance should be grudged. The simplest method is by means of a line, with marks placed at equal distances for the spaces between the plants up the hill. This carried by men at each end, who respectively measure a distance with a rod for that purpose from the last peg, and holding the line taught from end to end ; boys following with pegs ready cut put them in at the marks on the line. This method is likely to lead to inaccuracy on a large feature, in consequence of the irregularity of the ground, and from the small inaccuracies accumulating in the measurement between each laying down of the line. A better way is to lay out a feature in squares of the line, and these into parallels across the feature, and then drop the line between each peg on the parallels, putting in the remaining pegs by the marks on the line. Some planters have an excellent plan of laying down a number of lines up and down the hill at measured distances, and then two men with a shorter line measure distances upwards on the two outside lines, and peg-men put in their sticks at the points where the lines cross each other. We have seen some plantations laid out by the use of the theodolite, but this is not an instrument to be found on many estates. The object to be

E

gained in lining is to carry the plants in straight lines parallel to each other, at the same bearing over the whole estate, and to make them cross each other at right angles; by this means lines will be formed four ways, at the two right angles and at the two diagonals. When this has been successfully achieved, there is, under any circumstances, a workmanlike character about the plantation; vacancies in the planting are more easily detected, the arrangements for weeding, picking, pruning, and manuring are made without confusion. Diagonal lines across the features of land are said to be best, inasmuch as the stems of the trees by this arrangement offer somewhat more resistance to the washing down of soil by the rains.

Much argument has been held as to the proper distances at which coffee-trees should be planted; and in visiting the various districts it will be seen that all distances and all forms have been resorted to. The result of this extended experiment has produced a very general opinion in favour of close planting; that is, 6 feet by 6 feet, or 5 feet by 6 feet, or 4 feet 9 inches by 5 feet. The reasons in favour of close planting are these, that the extra number of plants to the acre is followed in the first two crops at least by a proportionate increase of crop. By close planting weeds are hindered from growing, and, what is of more consequence still, the ground is sooner covered by the growing plants, and therefore protected from the sun, exposure to which is found to impoverish more than anything else the surface soil, from which the plant chiefly seeks its nourishment. With an average good soil, and careful handling of the trees and manuring, it is found that closely-planted coffee may be made to continue to bear highly; that is, in proportion to the extra number of plants to the acre. Close planting is serviceable in enabling the plants to outlive the effect of exposure to high winds, which in some places are most destructive. The next proceeding is *holing* (which, as its name implies, is to make holes in preparation for planting the coffee, at the distances staked out by the lining). As coffee is planted in a virgin soil, which cannot be cleared of roots, rocks, and logs, for the operation of the plough, and were it cleared of these the disturbance of the soil would probably cause it to be carried away by the heavy rains; holes have to be made, and the larger these are, the nearer the approach is to that movement of the soil, which in general sets at liberty its fertilising properties. Holes are made from 18 inches to 2 feet every way, into which before planting the surface soil is scraped with a hoe. It is customary to contract for this work, which I believe is generally performed now with a sort of crowbar having a spud blade at one end; with these the roots of trees are cut through and rocks and stones taken out, the loosened soil being removed with the hand or with a cocoanut shell. This tool is furnished to the Singhalese because they prefer to sit down and work leisurely, but where men can be employed at day-wages, and provided with mattocks to break the soil and cut out the roots and stones, and hoes to clear out the loosened earth, the same work may be far more economically performed, inasmuch as the labourers being on their feet give not only their arms, but their whole body, to the exertion, and have not to raise themselves from a sitting position between each hole they have to make.

SECTION XI.

BUILDINGS, PLANTING, &c., IN CEYLON.

THESE comprise a bungalow for the manager, lines, or huts, for the labourers, and pulping-house and store for the preparation and reception of the crop.

Plate 5 shows the main works and buildings at Messrs. Worms' estates, Puselawa, Ceylon.

The position of the pulping-house should influence the selection of the spots for the other buildings, unless other circumstances determine it. The bungalow should, if possible, be an easy distance from the store, that the manager may without difficulty superintend the work, which at crop time is going on often at night as well as in the daytime. One set of lines for the labourers, for the same reason, should be near to the store, another set may be erected more centrically with reference to the field labour of the estate.

With reference to lines and bungalow, especially the latter, they will be erected according to the manager's taste, and in a great measure depend upon the material to be obtained in the particular locality for constructing them. True economy will be practised in making permanent and substantial dwellings which will require only slight repairs occasionally.

The selection of the land for the pulping-house and stores and drying-ground is all important, and should only be attempted by a resident manager to whom a long residence on the plantation has afforded an intimate acquaintance with every feature of the land. The advantages which have to be

Plate 5.—Main Works at Messrs. Worms' Estates, Puselawa, Ceylon.

Plate 6.—The Pulping House, Messrs. Worms' Estate, Pueelawa.

sought and combined are, availability of water power, both
to drive the machinery and for a water-course to the pulper,
on a site convenient to the fields from which the crop is to
be brought, and so level that it may be fitted for its purposes
at the least possible expense.

Water power is not always used, nor is it available on
every estate; but as there may be often good reasons to adopt
it afterwards, it should always be considered in the primary
arrangements. It may be applied to so many economic ob-
jects—saw-mills, drying apparatus, &c.—that it should always
be applied, if possible, to everything undertaken on the estate
requiring a moving power.

Plate 6 is an exterior view of a pulping-house on Messrs.
Worms' estates, Puselawa.

Coffee when gathered from the tree fully ripe is like a rich
scarlet cherry, out of which on being squeezed two coffee-
berries break forth, each covered with a light skin resembling
parchment, and moist, with a sweet mucilaginous fluid which
rapidly decomposes. The machine called a pulper is for the
purpose of removing the cherry skin or husk; this it does by
passing it between a barrel armed with perforated copper,
forming a grater, and a sharp-edged board called the chop,
which by means of wedges or screws is placed at the proper
distance from the barrel to ensure the greater part of the
coffee being pressed against it, but not so close that any of
the berries should be pricked through their parchment covering. As coffee is seldom uniform in size, much passes forward
from which the husk, or pulp, as it is called, has not been
completely removed; this, being separated by a sieve worked
by the machinery, is returned by hand to the hopper. The
coffee, deprived of its husk, goes forward by a channel prepared for it into a cistern, the pulps being thrown off behind;
these latter are now generally saved to be carried to the
manure pit. The coffee is left to soak in the cistern for the

night, or for that length of time which is sufficient to wash
off the mucilage, an operation which is facilitated by a slight
fermentation—not, however, always conducted with safety to
the future quality of the coffee. Pulping-houses are supplied
with several cisterns, into which the several pulpings may be
run off that the work may not stand still.

Plate 7 gives a view of the interior lower floor of the
pulping-house on Messrs. Worms' estates, Puselawa.

In the pulping-house above the pulpers is a large floor
called the cherry loft, into which the coffee in cherry from
the field is measured, and from whence, through holes, it is
made to fall into the pulpers, from the pulpers it is carried to
the cisterns, and when washed is taken out by the labourers
to be dried, which brings us to speak of the store and drying
apparatus. Before doing so we may mention that the pulper,
which is considered a very imperfect machine, is generally
driven by four men at two handles, aided by a fly-wheel; the
power is immediately applied to turn the barrel, to which is
connected a large cog-wheel, which moves a pinion attached,
to work a sieve or riddle.

Pulpers turned by water power, if properly erected to
resist the strain of the connecting machinery, work with more
equality of motion, and therefore do their work better than
those worked by hand.

Many improvements in the pulping-machine have been
suggested without success; the last, which is now very generally adopted, is called a crusher, and consists in a kind of
shield instead of a chop, which presses the coffee against the
barrel. This machine is said to do as much work in a given
time as the pulper, and with less liability to cut or prick the
berry.

It has been said that the pulper is an imperfect machine;
it is so in respect to the incompleteness with which it per-
forms the work for which it is constructed. The work com-

plete, would be to pass forward the whole of the cherry coffee from the hopper, as clean parchment coffee, into the cistern, and completely separate the pulp. Instead of this, not only is there much coffee passed forward unpulped, which has to be returned to the hopper, but much pulp comes forward with the coffee into the cistern. To overcome these imperfections, many hands are required attending the process of pulping; several with sieves receive all the coffee as it comes through the trough into the cistern, and remove all the pulp in it. The object of this is to please the sight by making the sample even, and to take away a material which both renders the coffee difficult to dry, and, being decomposed vegetable matter of a saccharine nature, is liable by fermentation to injure the quality of the coffee itself. The coffee which is passed unpulped, having been returned to the hopper as it accumulates at the end of the pulper riddle, until the picking has all been pulped, is sometimes passed through another pulper, set closer, and what then remains is dried separately in the husk. The imperfections of the pulper have recently attracted much notice, and improvements are making in it.

We now arrive at the crowning operation of all—*Planting*. Economy should be studied in every part of the work, and to plant to the best advantage, nothing is more conducive than having the traces of the roads and paths through the clearings already opened if possible, so as to be available for bringing the plants from the nurseries. Plants and stumps are both used in forming plantations; the first are seedlings, reared in nurseries until they are about 8 to 10 inches high, or just crowned—that is, after the appearance of their first lateral branches; the second are the stock and roots of an older tree. Under varying circumstances there is much to be said in favour of each. If the weather is dry and the season advanced, stumps may be planted with less risk, and in less time; but if plants are to be had, and the season favourable, it is generally considered they are the best. In putting them into the ground, if practicable, they should be removed from the nursery with balls of earth about their roots; this is even the most economical, as being almost an assurance against failure, and the tree thus planted receives no check, but begins to grow directly. If, on the other hand, the plants are to be brought too far to carry them with balls of earth, extreme care should be taken to place them with the tap root perpendicularly in the soil. Invariably the effect of a tap root not being placed perpendicularly in the hole is, that when the tree is grown to two or three feet in height, the upper shoot and branches take a paler colour to that of the healthy plant, which is of very dark green, the leaves also become small and elongated, and occasionally somewhat mottled with a yellow tinge. Such plants will frequently spring into flower and fruit prematurely, which generally turns out "boll," or empty in husk, and as prematurely dies away. This remark is not uncalled for, as it requires extreme vigilance to prevent the labourers carelessly doubling up the tap as they place the plant in the earth. The hole should be well filled up and the earth trodden in round the plant, which should never be buried below the crown of the root.

Nurseries of plants are variously made; much contention has existed on the merits of shade and no shade. Shade is not required, and the plants are best without it; yet it happens most nurseries are in some respects subject to shade, as they have to be formed frequently before any portion of the land is opened, and should be protected by a belt of forest from the fire of the clearing; the jungle being cleared off and carried to the sides of the ground selected. The soil is dug with hoes and picks about two spits deep, and the roots of the jungle carefully taken up, as these would afterwards break the rootlets of the plants when taken up for transplanting. The ground being laid out in beds, is sometimes sown broadcast with coffee-seed, which only requires to be most lightly covered with a little sifted mould, or the seeds are pressed in, in rows, with the finger. Sometimes small seedlings, with two leaves and the seed leaves, are brought from another nursery, or from beneath the coffee-bushes of a plantation or native garden, and lined out into beds about six inches apart; these make by far the best and hardiest plants for planting out into the fields.

We must not omit to notice the necessity for carefully replacing, with as little delay as possible, all failures in the planting. The longer this is delayed the more difficult does it become to do it well; and when neglected, besides making the fields appear irregular and unsightly, the gaps left become weed beds, and so much of the bearing space of the acre being lost, the crop is affected in proportion.

When stumps are planted, a number of buds or suckers make their appearance on the root; one of these only should be left, and the others carefully rubbed off with the finger, as often as the weeding party goes over the field.

The sprout from the bud which is left grows up and becomes the stem of a tree, and throws out its laterals somewhat higher than the tree grown from a nursery plant. The stump generally produces its first crop a little out of season; for planting up failures, stumps give the least trouble.

Next to lining and careful planting, nothing enhances the good appearance of a plantation so much as the workmanlike formation of the roads and buildings. It is, therefore, desirable, by the use of a theodolite, to trace the roads accurately before cutting them out. Formerly it was frequently the practice to take the road by the eye from point to point, evading the natural difficulties of rocks or large roots, by taking the breakneck path above or below them. Roads on partial clearings should always have reference to the land which is afterwards to be felled and connected with the same set of works.

Plate 8 shows a coffee district near Puselawa, Ceylon.

SECTION XII.

HARVESTING THE CROP AND PREPARATION FOR MARKET.

THE heavy blossom appears on the tree in August and September. The principal crop is picked from April to July. A small crop, chiefly from young coffee, is picked from September to December. The produce is sent down to Colombo, the shipping port, from April to September. If the estate be close to a carriage road this is done by carts, which can take from 60 to 80 bushels. The cost of transport is sometimes enormous. It is not unusual to see carts loaded with coffee lying at the bottom of a precipice, while the bullocks which had brought them have died from exhaustion. If not near a road, carrying coolies are employed, or pack bullocks, which take a load of 3 bushels, to transport it either to a store, from which a carriage will convey it to Colombo, or to the navigable point of one of the rivers. Of the various modes and facility, or the want of it, possessed by estates situated in different districts, some idea may be formed in the expense varying from 1s. to 12s. per cwt. for bringing the produce to Colombo.

There are now in Colombo upwards of thirty establishments for the preparation of coffee for shipment, ten or twelve of these employ steam power to drive the requisite machinery. To most of them large barbecues for drying are attached, and cooperages for the preparation of casks, and in the season, which lasts nearly three quarters of the year, from 10,000 to 15,000 women and 1000 to 2000 men are employed in the process.

Though the coffee has been sufficiently dried on the plantation to enable it to reach Colombo in safety, it is not sufficiently hard to part with the silver pellicle which envelops each berry under the parchment skin, and to resist the pressure of the peeler, without some additional drying in the more powerful sun at Colombo.

It is, therefore, again exposed on the barbecue, until it reaches a crisp dryness. (Plate 9 shows the barbecue or drying-floor on Messrs. Worms' estate, Puselawa, in Ceylon, and the native labourers spreading the coffee to dry.) It is next submitted to the pressure of the peeler, which breaks the berry out of the parchment covering, and sets the silver skin at liberty. It may be noticed that the silver skin, though perhaps not adding two ounces to the weight of 112 lbs., gives the coffee an appearance considered to be unsightly in the London market, and, therefore, depreciates its value; its adherence to the coffee, though the cause is not known in the market, is supposed to be generally the result of bad drying on the plantation, being allowed to remain too long wet, or being permitted to heat after it is taken from the cisterns.

Several processes have to be gone through before the article known in commerce as coffee is produced. In the first place, the pulpy exterior of the berry, as we have seen, has to be removed by the process of pulping, which separates the seed and its thin covering, called the parchment, from the husk. When this pulping process is completed, we have the parchment by itself in a cistern, and the next process consists in getting rid of the mucilage with which it is covered. For this purpose the water is drained from the cistern, and fermentation is allowed to take place, which it readily does after a period of twenty-four hours, or even less on a low estate, where the climate is warm, though forty-eight hours are generally required on the highest estates.

Plate 9.—The Barbecue, or Drying-Floor, Messrs. Worms' Estates, Puselawa.

Water is then admitted into the cistern, and the coffee being agitated by wooden rakes, the mucilage combines with the water and is drained off. After this the washed parchment coffee has to be dried to a hard stage, and as it frequently happens that during crop time there is a continuance of wet weather for weeks and months together, the chief difficulties which a planter has to contend with now present themselves.

The peeler, or machine for removing the parchment, consists of a circular trough, in which a wheel is made to travel; this is generally made of wood, and shod with copper sheeting, and is turned by central pressure, like the capstan of a ship, either by hand or by the gearing of machinery attached to a steam-engine. An improvement on this has been made by constructing the travelling wheels of iron, and the trough of plates of the same metal; these plates being serrated in one direction, so as to present a rough surface to the coffee, facilitate the fracture of the parchment. Two wheels are generally made to work in one trough, each of which is provided with a kind of scraper, to stir up the coffee in its path and cause it to present a new face to the pressure.

A coffee-peeler is usually made of durable wood or iron. The circumference of the machine is 36 feet; the breadth between the circles in the machine is 1 foot. The height of the wheel 6 feet; the thickness near the axle-tree 1 foot, and on the top 6 inches; twelve men or four bullocks can turn it. If turned by men 200 bushels of coffee, and if worked by bullocks 140 bushels, can be obtained in nine hours; if by steam about 800 or 1000 bushels. The cost of constructing a machine to be worked by men or bullocks is 25l., by steam 600l.

After undergoing this process, the coffee is passed into a winnower, which removes nearly the whole of the parchment and silver skin. It is now given to the women, in quantities of a bushel each, to be picked over by hand, who take out all blacks, broken berries, *triage*, or anything calculated to injure its even quality. Further to improve its appearance, the coffee is passed through a sizing machine, by which generally three sizes are separated : the round, or pea berry, and a larger and smaller berry, each of which from the separation is more even in appearance, and as such preferred in the London market.

Sizers are variously made of perforated sheet zinc or wire gauze, with openings of three sizes, increasing from the top in the form of a long pipe, which, being slightly inclined and made to revolve, pass the coffee poured in above into the bins constructed to receive the different sizes.

The coffee, now ready to be packed, is at once put into casks, containing 6 or 7 cwt. each, and sent away on board ship without delay. These processes are constantly improving and are now thoroughly understood—a remark which would seem uncalled for, but for the recollection of the bungling and conflicting systems which were in vogue a few years since.

SECTION XIII.

PREPARATION FOR MARKET—(*Continued*).

PARCHMENT COFFEE, when in an unseasoned state, is prone to enter into decomposition from the time at which it is withdrawn from the protection of the living organism until it is thoroughly seasoned by drying, after which it may be kept for any length of time in a dry place.

The worth of coffee as an article of commerce is lessened in proportion to the extent to which these progressive changes are allowed to go on. If heating has taken place, the bean can never afterwards acquire the pellucid colour which is indicative of well-dried coffee, but partakes more or less of a dingy appearance. If mouldiness ensues, the aromatic properties, like those of tea, give place to an insipid flavour; and finally, if the bean undergoes putrefaction, it assumes a dull black colour, and becomes totally destitute of every valuable property.

When the crops ripen, they must be gathered and cured under all circumstances of weather; and as it generally happens that this has to be done during the prevalence of the periodical rains, the difficulties to be contended with are so much the greater. The extensive nature of the operations has also to be taken into account in forming an estimate of the difficulties to be provided for. During the busy season of crop upwards of 1000 bushels of cherries are daily gathered from some plantations, yielding an increase of 500 bushels of parchment coffee to be daily added to that which has already accumulated in the store.

Past experience having shown that coffee was most easily

preserved in a sweet state when spread thinly on the floor, large and commodious buildings were called into use, notwithstanding the unusually heavy expense which attended their erection in situations remote from town, where sufficiently skilful labour was only to be had at the time, and with great difficulty. On this account inadequate accommodation was provided on many plantations, and the coffee accumulating to a considerable depth, no amount of hand-turning could keep it from contracting a musty smell, its proneness to decomposition increasing greatly in proportion to the extent of the accumulation.

Some years ago it occurred to Mr. Clerihew that it was possible, by means of fanners, working on the exhausting principle, so to withdraw air from an enclosed space as to establish a current of air through masses of coffee spread on perforated floors forming the top and bottom of that space. This plan he carried into execution at Rathoongodde plantation, and it has since been adopted by many planters.

The following is a detailed description of Mr. Clerihew's invention, a model of which was shown at the International Exhibition of 1851:

The water-wheel is an overshot one, nine feet in diameter, and is of much smaller dimensions than any wheel that has hitherto been employed for pulping. It is, however, sufficient in power to work the fans and pulpers simultaneously, the excess of its power over that of other wheels being gained by the diminution of friction consequent on there being no intervening shafting and gearing. The entire wheel is constructed of wood, with the exception of the shaft, which is unusually light, as it has merely to serve as a support to the wheel. By means of a double band rim-bolted to the arms on each side, motion is given to the pulpers from the one and to the fans from the other.

The floors of the curing-house are laid with laths $1\frac{1}{4}$ inch square and 2 inches apart; these are covered with open coir matting; being rather cheap and durable, this material answers the purpose remarkably well. The side walls of the curing-house are constructed in the manner of the country, viz. of wattled work filled in with clay and smoothed over so as to be air-tight.

To derive the full benefit of natural heat, the roof is covered with felt or with sheet-iron, so that in fine weather the temperature of the air in the upper floor is raised considerably by contact with the hot roof, and its capacity for absorbing moisture much increased, preparatory to its being drawn down through the mass of coffee in the upper floor. Even in the cool climate of the district of Upper Hewahette, at an elevation of 4500 feet, in a fine day the temperature of the air under a felt roof is 120° when the fans are not working, so that a great drying power is thus made available at no expense.

The lower floor, on the other hand, is adapted for the application of artificial heat for the purpose of evaporating the surface water from each separate batch of coffee as it is taken from the washing cisterns preparatory to its being deposited in the upper floor. In wet weather this is essential, for the atmospheric air being then saturated with moisture, no drying can take place until its capacity for absorbing moisture is increased by an increase of temperature. 'One other reason for adopting this arrangement is, that when the coffee is first taken wet from the washing cisterns the interstices of the beans are more or less occupied with water, and thus present a medium less pervious to air than is the case when the surface water has been dried off. Consequently it is desirable that, until this has been done, the depth of the coffee should not exceed six inches, and, to be equal to every emergency, the heating power ought to be sufficient in the wettest weather to evaporate the surface water from the produce of a day's picking (within twenty-four hours), so as to

F

allow of its being removed into the upper floor. The daily number of bushels picked from any, or the same plantation, is of course a variable quantity : depending on the extent of the cultivation, the quality of the trees, the number of hands employed, and the elevation of the land ; the latter, when considerable, having the effect of prolonging the picking season. The stove is more than sufficient for a daily picking of 400 bushels of cherries.

The heating stove is square, has a waggon head with a semicircular opening in the centre for the passage of air, and is constructed of stout sheet-iron. It is placed within an arch, with a clearance of nine inches all round also for the passage of air, the guiding principle in its construction being to adapt it to the burning of wood, and to expose as much heating surface as possible to the air which flows past it into the air-chamber beneath the ground-floor. The stove opening is the only one which admits air to the coffee on the ground-floor. Consequently, when it is more or less closed by a damper, the power of the fans is exerted either in part or altogether on the mass of coffee in the upper floor.

In these applications of natural and artificial heat to the curing of coffee, the heat is conveyed by the air through the whole depth of coffee in such a manner that each bean feels its influence, whilst the watery products elicited by the heat are at the same time, and by the same means, carried off. It cannot be doubted that these applications are far more effectual than any of the modes hitherto in use; in some cases stoves were employed in the apartment containing the coffee, but it is obvious that their influence could not extend beyond the surface of the mass, and that, if the apartment was closed, there was no provision for carrying off the air that had become loaded with moisture due to its temperature; whilst, if the apartment was open, so as to afford a free draught of air, the greater portion of the heat given out by the stove was carried out before the heated air could act on the coffee. In other cases, heating pipes of various kinds were used below the floors on which the coffee was placed. This arrangement, however, has the effect of injuring the coffee, by steaming it; no provision being made for carrying off the excess of hot watery vapour which accumulates within the mass, but, on the other hand, the natural processes of decomposition are assisted and promoted.

The construction of Mr. Clerihew's heating apparatus is simple, and a moderate supply of fuel has a considerable effect in raising the temperature of cold damp air before it is brought into contact with the coffee through which it is drawn by the aid of the fans. This heated air becomes diffused throughout the whole of the chamber, which extends beneath the ground-floor in such a manner that no portion of the coffee which is on that floor can be free from its influence.

The fans at Rathoongodde are much more powerful than those in common use, the peculiarity in the shape of the blade giving them a great advantage as air-moving machines, in so far as the indraught is concerned, whilst one-half of the periphery being open a ready exit is afforded for the discharge of air. In the ordinary fan, if a smoking match is applied to any part of the indraught opening, the air will be seen to flow towards a neutral point in the centre of the fan, following a spiral direction, and thence in the periphery of the fan.

In Mr. Clerihew's modification of the blade each film of air, so to speak, flows into the fan directly, until it impinges on the curvilinear part of the blade, and from that point is thrown at a right angle towards the periphery. The column of air being thus less distorted in its progress, there is not only a greater quantity discharged, but much less power is consumed in effecting that discharge—in the common fan it is evident, from the circumstance of the air flowing to an apex, that a great amount of power is wasted in producing the increased velocity with which a column of air equal in volume to the two ingress openings of the fan must pass so contracted an area before it is discharged; hence it is that the fan, as an air-moving machine, has been considered unequal to the screw.

The enclosed space of the coffee-curing house at Rathoongodde has an area in the cross section of 100 superficial feet, it is 70 feet long, and a pair of fans are placed at one end. Repeated experiments have shown that, when the fans make 100 revolutions per minute, a cloud of smoke travels to them from the centre of the enclosed space (a distance of 35 feet) in precisely 15 seconds, hence we have $100 \times 35 = 3500$ cubic feet of air discharged in a quarter of a minute, or 14,000 cubic feet per minute; a screw of nearly seven feet in diameter would be required to discharge the same amount of air, and the cost of it in England is 84 guineas, whilst the pair of fans made and fitted up at Rathoongodde cost under 9l.

In the centre of the enclosed space, with a depth of four feet of coffee in the upper floor, the flame of a candle is blown to a right angle when the whole power of the fans is put on that floor; near to the fans it is extinguished, the air moving forward with a uniformly accelerated velocity from the farther end towards the fans, owing to the constant accessions made by the air entering the enclosed space throughout its whole length.

It has already been mentioned that the only entrance of air into the air-chamber beneath the ground-floor is by the opening in which the stove is placed, consequently, when this opening is closed by a damper, it is obvious that the whole power of the fans is exerted on the mass of coffee which is being cured on the upper floor, and that the division of this power may be regulated at will by more or less obstructing the entrance of air to the air-chamber by the damper. The upper floor is not supposed to be an air-tight apartment, but as the chief entrance of air is by the two doors in the end, its influx may be so far obstructed by closing them as to throw the greater part of the power of the fans on the coffee which is on the ground-floor, when this is required. Again, since it is obvious that, in wet weather, when the atmosphere is fully saturated with moisture proportionate to its temperature, it becomes a desideratum to introduce a portion of the artificially heated air into the vacant space which is over the coffee in the upper floor, so that the air which passes down through that coffee may have an absorbing tendency; this is accomplished by shutting the doors of the upper floor and throwing open the top-covering of the fan. By this means one-half of the air which is drawn from the stove is thrown in above the coffee in the upper floor, whilst the other half is discharged altogether. This infusion of heated air would on many occasions be attended with benefit, but the advantage will naturally depend on the comparative state of dryness of the coffee on the two floors.

In having recourse to these practical modifications some little judgment and observation are of more service than precept. It will be found, for instance, that if the coffee in the upper floor approaches the dry stage, it is better in wet weather to shut the doors of that floor as well as the tops of the fans, so that only a small flittering of air sufficient to ward off the first stages of decomposition may pass through that coffee, whilst the wet coffee below has the full benefit of a more rapid circulation of absorbent air.

Attention may now be directed to the practical results which these arrangements have afforded in the curing of coffee.

The coffee in the upper floor, as the crop advanced, gra-

dually increased in depth until it stood at four feet all over
the floor. When at this depth, with the fans making 100
revolutions per minute, the flow of air was quite sensible to
the hand placed on the surface of the coffee, and was rendered apparent by the smoke from a match following the
direction of the air; at the same time the rarefaction of the
air within the enclosed space was so very slight as barely to be
appreciable by a very delicate mountain barometer, though it
had the effect of causing the door to shut with a slam; thus
showing that a slight rarefaction of the air is sufficient to
disturb the balance of atmospheric pressure, even when acting through a medium of coffee of considerable depth. The
current of air thus established continued to flow without in-
terruption until the fans were stopped.

A cold glass tumbler taken into the store in a warm day,
when the fans were not in motion, became instantly dimmed
and wet by the precipitation of moisture from the internal
air. When another glass was taken into the store, one
minute after the fans were put in motion, it [remained clear,
without a trace of moisture.

A very satisfactory result soon showed itself, viz. that
whilst the temperature of the air as it entered the moist
coffee was 80° in a warm day, the temperature of the coffee
itself, as indicated by an immersed thermometer, was only
58°; the wet coffee being invariably coldest when the air
that was made to pass through it was warmest. This para-
dox admitted of easy explanation, when it was considered
that each bean of undried coffee was under similar circumstances to evaporating vessels of water placed in a draught of
warm air for the purpose of cooling the water. The cold thus
produced was, therefore, the necessary concomitant of the
evaporation that was going on, and the difference between
these temperatures afforded a measure of the drying power
in different states of the weather. The circumstance of the

hot air lowering the temperature of the coffee was also favourable in another point of view, seeing that it has been shown that heat is one of the conditions which promotes mouldiness, or the germination of fungi.

Every bushel of parchment coffee contains half a cubic foot of air, a fact ascertained by a bushel which took thirty-three measures of water to fill it. When full of newly-washed coffee it took thirteen of these measures of the water to displace the air from the interstices of the beans without overflowing, so that we have 13—33 of air in a bushel of coffee ; in other words, half a cubic foot. Hence, the fans in use are capable of giving a fresh atmosphere to 28,000 bushels of coffee every minute, or in the same time four fresh atmospheres to 7000 bushels.

During a continuance of nearly three months of wet weather which occurred at one crop time, the coffee in the curing-house dried very slowly, but was kept in a perfectly fresh and sweet state without the intervention of any manual labour, further than in depositing each day's increase in the lower floor to dry off the surface water, and removing that of each previous day to the upper floor, where it was spread on the top of all the coffee that had previously accumulated. These three months of wet weather were succeeded by a fortnight of very dry weather, and, on examining the coffee at the end of that period, it was found to have reached the dry horny stage at which it is usual to despatch it from the estate to Colombo, for the purpose of being peeled and shipped. On examining the beans they were found to be of that clear colour which distinguishes coffee carefully cured in small quantities, with the advantage of the most favourable weather. Under like circumstances, viz. during such a continuation of wet weather, it would have been impossible to preserve the coffee free from more or less mustiness of smell, by manual labour employed in the usual way, whilst at th same time the expense of storework would have been more than fourfold. The whole expense of the storework, viz. pulping, washing, curing, and storing the Rathoongodde coffee, amounted to 2¼d. per cwt., and when it is considered that during crop time the value of every man's labour is greatly increased, it is an object, as far as possible, to substitute mechanical contrivance for manual labour, so that all hands may be employed in gathering the crop as it ripens.

Every planter knows that when coffee is spread out in a single layer on the floor of his store, it becomes dry after a time, and is well cured without any further attention on his part; but it is impossible to devote sufficient space for this purpose without incurring an expense which would be quite incompatible with his circumstances. When, however, coffee is thus spread out in a single layer, it is obvious that the reason why it requires no attention is, because the beans being freely exposed to the atmosphere, there is naturally a constant change of the air by which they are surrounded; the same air is not sufficiently long in contact with the beans to excite the first action of decomposition, and the absorption of oxygen is not accomplished. Presuming, however, that it were so, the subsequent actions could not take place, for the products of the first action, viz. carbonic acid, heat and watery vapour, would immediately make their escape and be dissipated by the atmosphere, which is precisely what takes place when, by mechanical means, a draught of air is carried through a mass of coffee. Hence, it is evident that the requirements of space are overcome by the adoption of this plan, and that a great mass of coffee is placed under conditions similar to those by which a single layer is influenced when exposed to a natural draught of atmospheric air.

SECTION XIV.

CULTIVATION IN SOUTHERN INDIA.

SOUTHERN INDIA is becoming as celebrated for its coffee, as Northern India for its tea. We find that the exports of coffee from Madras have increased considerably during the last five years, and there is every reason for supposing that Southern India will shortly become the chief coffee-producing country of the world. We have no idea of the number of acres of land under coffee cultivation in the Madras Presidency, but it must be very large, for after its local wants have been supplied, coffee to the value of half a million sterling is exported.

In 1858-59 the shipments were 7,288,421 lbs. to foreign ports, and 4,083,917 lbs. to Indian ports. In 1862-63 the shipments were 16,292,238 lbs. to foreign ports, and 3,976,766 lbs. to Indian ports.

Though some parts of India are well adapted to the culture, it is not yet so extensively cultivated as might have been expected from the vicinity of its Arabian sites to the Malabar coast. There, however, some excellent coffee is grown, as well as in the hilly regions of Mysore and on the slopes of the Neilgherries, and some of these are of such good quality, and so carefully prepared, as to bring the same price as Mocha coffee. Some very good specimens of coffee have also been produced in the interior of India, as in the district of Chota Nagpore, where the culture might apparently be greatly extended, and be of great benefit for consumption in that part of the country.

According to local tradition, the coffee-plant was introduced into Mysore by a Mussulman pilgrim, named Baba
Booden, who came from Arabia about two hundred years
ago, and took up his abode as a hermit in the uninha-
bited hills in the Nuggur Division named after him, and
where he established a college, which still exists, endowed by
government. It is said that he brought some coffee-berries
from Mocha, which he planted near to his hermitage, about
which there are now to be seen some very old coffee-trees.
However this may be, there is no doubt that the coffee-plant
has been known in that neighbourhood from time imme-
morial, but the berry has never come into general use among
the people for a beverage. It is only of late years that the
coffee trade of these districts has become of any magnitude,
or that planting has been carried to any important extent.
The export of coffee from British India, which in 1851 was
only 3239 tons, had increased in 1861 to 8535 tons; about
one-fourth of this is shipped from Bombay, and nearly all
the remainder from Madras.

More than thirty years ago a few Europeans were engaged
in coffee planting near Chickmoogloor, a few miles from the
Bababooden Hills. About twenty years ago the plantation
producing the well-known coffee called "Cannon's Mysore,"
and others, on the Memzera, or "Bad Mountain," was commenced by two enterprising gentlemen. The success of
these has induced many more Europeans to plant coffee there,
and the consequence is that the coffee trade of Mysore
bids fair to emulate that of Ceylon. It has given, also, an
example to other parts of India, and the plant originally
taken from the Bababooden Muth is now extending over tens
of thousands of acres in Coorg, the Wynaad district, the
Neilgherry Hills, and along the Western Ghauts, north and
south.

In Mysore the number of European coffee-planters has increased to about thirty, while the number of native planters is estimated at between three and four thousand.

The average produce per acre in Mysore is probably not half that of Ceylon. Some attempts have been made to cultivate coffee in the open country, but without success; it seems to require forest land and considerable elevation and moisture. " Cannon's Mysore" is grown on a range of hills from 3500 to 4000 feet above the sea, having the benefit of the south-west monsoon, which very seldom fails at all, never entirely, and of the tail-end of the north-east monsoon. This elevation gives a pleasant climate, well suited to Europeans.

Several species of the genus Coffea (*C. alpestris*, *C. grumeloides*, and *C. Wightiana*) are indigenous to the Neilgherry Hills.

A berry, generally one which has itself fallen ripe from the tree, is put into the ground, usually in a nursery plot, though some planters prefer to place the seed in the identical hole which is to be its future situation. The nursery plan is, however, generally adopted, and here the young plant, which shoots up in about a month after it is sown, is allowed to remain until about sixteen months old. It, or rather we will say they, for hundreds and thousands are generally dealt with at once, are transplanted to holes which have been carefully prepared for them on the soil which is to be their future location. These holes are generally two feet cube, and many good planters prefer them even deeper; in this the plant is carefully placed and covered around, and in eighteen months from that time, *i.e.* about three years from the time the berry was first planted, our small coffee-tree begins to bear fruit, the first crop being of course very scanty.

The berry is picked from November to the end of February, by any number of men, women, and boys which can be collected, and who are paid by the quantity they pick, some expert hands earning a good deal.

The berry collected is carried to the house of the estate, and there having been weighed, is thrown into what is called a cherry loft, a wooden chamber, alongside of, but a little higher, than the place containing the pulper. From this cherry loft to the pulper the coffee is washed by a stream of water, which carries it along a trough so arranged as to catch and impede any stones or heavier materials from entering the pulping machine. These heavier materials sink to the bottom of the trough, and the buoyant coffee-berry, floating on the surface, is borne to its destination.

The object of the pulper is to remove the fleshy capsule from the berry, and this being accomplished, the coffee passes on in one direction, whilst the pulp, by a clever arrangement of the mechanism of the instrument, is pushed away in another. The berry is now thrown into a vat and allowed to ferment, until the remaining mucilaginous substance adherent to the parchment covering is easily washed away by water.

This accomplished, it is thrown on open exposed places, called barbecues, and allowed to dry in the sun. This takes about twelve days, when it is packed in gunny (jute) bags, placed upon bullocks, and despatched to the coast.

There it is what is called garbled, that is, having been once more exposed to the sun and thoroughly dried, it is placed in circular troughs, and over it large heavy wheels, shod with iron nails, are made to revolve. This removes what is called the parchment skin, leaving the berry now covered only with a beautifully fine coating, the silver skin.

It is then, by a number of women employed for the purpose, carefully sized; after this, passed through a pea-berry mill, the object being to separate the round pea-shaped berry from the flatter, the former being much more prized, and fetching a higher price in the market, though why, it is difficult to say, as it makes no better coffee than the other; and as it has to be deprived of its form by roasting and grinding before it comes to table, the advantage of its pea-shaped figure is, to say the least of it, somewhat obscure.

There is likewise separated from the rest what is called "triage," the broken and otherwise defective beans, which are also packed by themselves, and which again, we believe, though selling more cheaply, are found to make quite as good an infusion for a beverage as their more aristocratic friends the pea-berries. However, pea-berries, flats, and triage, are all ultimately packed in square wooden boxes and shipped to England, where it is sold, roasted, ground, and drank.

In Wynaad, in the close of 1863, there were 93 coffee estates, covering 50,000 acres, of which about 15,000 acres were planted; 6100 acres had trees over two years old on them. There were also about 3600 acres under culture with coffee by the natives. Wynaad is an elevated plateau, rising somewhat abruptly from the western or Malabar side, but sloping more towards the Mysore or easterly side.

The quantity of coffee exported from Tellicherry during the official year ending April, 1862, was 58,500 cwt., of which about 30,000 cwt. is supposed to have come from Wynaad, the rest from Coorg. 8 cwt. of coffee per acre is considered an average yield in Wynaad, 10 cwt. a good crop.

SECTION XV.

BOURBON, JAVA, AND THE EAST.

It was from Beit-el-Faguil, the European factory near Mocha, that the coffee-tree was transported to the island of Bourbon, in the year 1718, and it is remarkable that the islanders recognised the plant as natural to their own country, and brought the astonished importers abundance from their native mountains. In Bourbon they distinguish four varieties of the coffee-plant.

1. The Mocha, which is very delicate, for the plants degenerate and often perish after a good crop.

2. The Levoy, which is more hardy, but the coffee is inferior in quality.

3. The Myrtle, a variety of the Mocha, very hardy, and yielding abundant crops.

4. The Marron, or wild coffee, with such bitter and narcotic properties that it can only be used by admixture with the berries of one of the other varieties.

JAVA.—In Java, coffee is a government monopoly, and the planters bring their coffee to a central government depôt for sale at a fixed price. The island exports about 1,250,000 cwts. of coffee annually. Java coffee has lost much of its former repute from being largely saturated with moisture, artificially to the extent of 14 per cent.; this increases the weight, but must injure the quality in transport.

At the Paris Exhibition of 1855, the Netherlands Commercial Association contributed a very varied collection of two dozen varieties of coffees from the Dutch government possessions in Java, under the following classification: Brown, clear brown, deep yellow, yellow, yellowish, white,

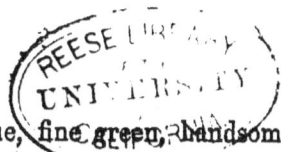

whitish, pale of Havana kind, blue, fine green, handsome
green, green, greenish, mottled green, deep green West India
kind, green West India kind, pale green West India kind,
dark Demerara kind, green Demerara kind, deep grey, triage,
common black, greenish Menado, and white Padang. The
Netherlands Society sell about 1,000,000 bags or bales of
coffee annually.

The following were the exports of coffee from Java in
1862:

	piculs.
To Holland on private account . .	128,047
To other countries	165,116
By the Netherlands Trading Company .	877,241
	1,170,404

The position of the coffee trade of Java is shown in the
figures annexed, for five years :

EXPORT AND VALUE OF COFFEE.

	tons.	value.
1858	66,575	. £2,614,505
1859	59,769	. 2,565,137
1860	54,638	. 2,486,115
1861	61,783	. 2,850,518
1862	63,286	. 3,465,747

Three kinds of Java coffee are commonly brought to
Europe—Jacatra (usually sold as Java), Cheribou, and
Samarang. The first is the best, the second is generally a
little lighter colour and of somewhat inferior quality, and
the third has yellowish brown, or green, flattened beans.
What is generally sold in the Dutch markets as Samarang
is, however, simply a kind of "triage," with black beans of
a coarse flavour.

SIAM.—On the hilly districts of the east coast of the Gulf
of Siam the cultivation of coffee is carried on to a limited ex-

tent, and some very fine samples of Siam coffee were shown at the International Exhibition of 1862, sent me by Messrs. Markwold and Co., and by Sir Robert Schomburgk, the British Consul-General.

SUMATRA is one of the worst kinds of coffee received from the Eastern Archipelago. The beans are large, dark yellow or brown, and occasionally even black, and the flavour varies considerably. The production in Sumatra averages about 5 to 6,000,000 lbs., but has often been double that amount.

CELEBES.—With the exception of Menado, which has large beans of a pale greenish or yellow colour, Celebes coffee is greatly inferior to Java, and it is questionable whether the colour when brought to market is not given by artificial means. The production is about 1,000,000 lbs.

PHILIPPINES.—Manilla coffee is one of the best of the Eastern kinds, and quite equal to Java. The average production is about 3,000,000 lbs. The beans are medium-sized, and of a pale greenish colour. The coffee is shipped in bags of about 150 lbs., or in cases or chests of 200 lbs. to 300 lbs.

OTHER SOURCES.—The cultivation of coffee is making rapid progress in the Sandwich Islands. There are now considerably more than half a million trees in bearing on the island, producing upwards of 2,000,000 lbs. annually— the largest proportion of which is shipped to California. Queensland and the northern districts of Australia could raise large quantities of coffee. It is much less laborious than cotton, more fitted for women and children, and, being adapted to the mountain ranges of tropical climates, of course more healthy and invigorating than the sultry plains. The range of mountains varying from twenty-five to thirty miles from the northern coast of Australia, towards Torres Straits, would be admirably suited to the culture.

SECTION XVI.

COFFEE AS A BEVERAGE.

IT is remarkable that, much as coffee is used in this country, the proper mode of preparing it as a beverage should be so little understood. Perhaps it is that most people consider coffee-making as too easy a process to need any pains at all; and for this reason the coffee served at nine breakfast-tables out of ten, throughout the kingdom, is a miserable muddy infusion, which people seem to drink only because, as washerwomen say, it is "wet and warm." The right way of making coffee is not less easy than the wrong one; there is no mystery about it. All that is required is the observance of a few simple rules:

1. The nature of coffee is such that it parts very easily with its aromatic, stimulating, and other properties; a small quantity of water will draw out all the goodness quite as effectually as a large quantity, and it will do this if the coffee-berries be only bruised or very coarsely ground. It is a grave mistake to suppose that coffee should be ground to a fine powder; extreme fineness is the great cause of "thick coffee" as prepared for breakfast. In Eastern countries, where people know what good coffee means, they always bruise the berries in a mortar. In fact, the goodness of coffee depends more on the roasting and the method of preparing afterwards, than on the quality of the berry, or any other particular.

2. Buy your coffee ready roasted, but not ground; that is, buy coffee-berries, and always choose such as are fresh

G

roasted, in preference to stale. Observe, also, whether your grocer keeps the article properly shut up in tin canisters, or lets it lie about in open tubs or trays.

3. If possible, buy a coffee-mill, one that will grind very coarsely. The price varies from 2s. 6d. to 5s. This article is so essential to a good cup of coffee, that no one who can afford the outlay should hesitate to buy one. Those who have a pestle and mortar may try the method of bruising ; ,but whether a mill or a mortar, no more should be ground or crushed than is wanted for use at the time.

4. Coffee requires to be kept in a very dry place; and, as it readily takes up the flavour of other articles near which it may be placed, it should be kept in an air-tight vessel. If you buy tea and coffee at the same time, do not pack them in one parcel or basket, or carry them in the same packet, for the true flavour of both will be injured. We presume that no one will be so careless as to keep either tea or coffee in paper only ; a wooden box would be better than this, but a bottle or porcelain jar is best of all.

5. Have a clean, dry coffee-pot; it should always be rinsed out when put away, and turned down to drain.

6. To every half pint of water, allow half an ounce of coffee-powder; have your kettle of water boiling, put the necessary quantity of powder into the coffee-pot, and pour in as much water from the kettle as you require. Set the pot on the fire for a few seconds, but on no account let the contents boil up ; then pour about half a pint of the liquor into a cup, and pour it back again into the pot, and stand it on the hob or on the fender to settle. If these directions have been properly followed, there will be in three or four minutes a pot of coffee as clear and well-tasted as any one could wish to drink. Should it be too strong, you have only to use less of the coffee-powder. All the goodness is extracted with the first boiling ; and those who wish to drink

good coffee must never boil the same grounds a second time.

7. The milk in all cases must be warmed, and used as hot as possible; and it should always be put into the cup with the sugar before the coffee is poured in. When a cup of coffee is taken after dinner, it should be drank without milk, and with little or no sugar.

8. But of all the preparations of coffee there is none equal to the French, known as *café au lait*, or milk coffee. We have drank it constantly for several years, and can pronounce it to excel all others as a breakfast beverage. In this there is more milk than water, and the coffee liquor is rather an essence than a decoction; it will be almost black in colour. The process to be followed is the same in most respects as described; but, instead of a quart or three pints, not more than a third of your usual quantity of water is to be poured on the full quantity of coffee-powder. After it has stood to settle, pour it carefully off the grounds into a jug or pitcher, which is to be kept hot by any convenient means. In this way the liquor, though black, will be perfectly clear. At the same time a quantity of milk, according to the wants of your party, must be heated in a saucepan with a spout or a lip. When this is ready, pour it into your breakfast-cups until they are three-parts full, or rather more, add the sugar, and then fill up with coffee from the jug, more or less, according as you prefer it strong or weak. Coffee made in this way will be found more nutritious, and to possess greater richness and smoothness, than can be attained by any other means.

Many persons are in the habit of keeping roasted coffee in vessels of tin, closely secured; this is a most improper mode, and the consequences of doing so may be pointed out. It is known that coffee contains gallic acid, a principle which has the property of acting on iron or tin, and it is therefore certain that in keeping coffee in these canisters the acid has such an effect in dissolving particles of the metal, as not only to affect the taste, but even the colour of the coffee. To convince one's self of this, it is only necessary to leave some freshly-roasted coffee in a tinned vessel for a time, and it will soon be found to have imbibed a black colour and a most disagreeable taste.

It appears, therefore, to be necessary to avoid keeping coffee in these metal receptacles. The best mode of properly preserving the article is by using vessels of porcelain, or other similar material. With regard to the description of coffee-pot to be used in preparing the article, it should never be of tin or iron; nothing will so soon and so surely destroy the fine flavour of the beverage as these descriptions of coffee-pots. It has generally been the practice to make coffee either by boiling it, or by pouring boiling water on the ground coffee placed on a filter. Both of these methods are bad.

Experience has shown that boiling water destroys or sensibly alters the volatile parts of the berry, and dissolves those which are bitter and unpleasant. We ought not, therefore, to employ water heated to a greater temperature than to allow the finger being placed in it. But difficult as it may be to believe, there can be no doubt but that the best mode of preparing this beverage is with cold water. Coffee so made is not only more aromatic, more limpid, and more substantial, but it is far stronger than any made with hot water. The cold infusion takes from the coffee and communicates to the water all its aromatic qualities, while it does not imbibe much, if any, of the gallic acid; consequently this preparation is far less bitter than that which has been boiled, in which process the most minute particles are acted upon.

Coffee thus made is of a fine bright and dark colour; it requires far less sugar and much less care, because all that has to be done is to place the powder on the filter, drop on it a little water, and when well moistened to pour on it the proper quantity of water. The filtration will be completed in a moderately short space of time, and the liquor having run through, may be again poured on the coffee, so as to remove any further portion of flavour left in it; and when this has been done, the preparation will be so delicate and aromatic that those who taste it will adopt the mode in preference to any other. When the coffee thus made is to be warmed for use, it must not be heated to the boiling point, and take care that the vessel in which it is warmed be quite full. It may be here remarked that coffee thus made warm is always more pleasant than when drank at the time of its preparation, provided it be not made to boil, and that the coffee-pot be well closed. It is equally necessary with the above that the berry should be well and thoroughly roasted, and not ground in a mill or machine, but pounded and sifted, so as to secure the particles being of equal fineness.

To enter into an examination of the comparative merits and demerits of the several percolators and cafetières at present in use, would extend these observations to too great a length; but most of those generally adopted are worthless, or complicated, with the abominable bag-filter, which is seldom kept clean. There is ample room for inventors in the manufacture of a simple coffee-pot with a water-gauge at the side, which shall effect what is not now done—a passage of the hot water once only through the coffee, so as to have a bright infusion instead of a muddy decoction.

"Tea," observes Dr. Sigmond, "as the morning beverage, when breakfast forms a good substantial meal, upon which the powers for the day of meeting the various chances and changes of life depend, provided it be not strong, is much to be recommended; but when individuals eat little, coffee certainly supports them in a more decided manner; and, besides this, tea without a certain quantity of solid aliment, is much more likely to influence the nervous system. Some persons, if they drink tea in the morning and coffee at night, suffer much in animal spirits and in power of enjoyment of the pleasure of society; but if they reverse the system, and take coffee in the morning and tea at night, they reap benefit from the change; for the coffee, which to them in the morning is nutrition, becomes a stimulus at night; and the tea, which acts as a dilutent at night, gives nothing for support during the day."

The Turks drink their coffee very hot and strong, and without sugar; occasionally they put in, when boiling, a clove or two bruised, or a few seeds of staranise, or a drop of essence of amber.

The following quotations from recent travellers give the Turkish mode of making coffee :

"The bruised or ground beans are thrown into a small brass or copper saucepan; sufficient water, scalding hot, is poured upon them, and, after being allowed to simmer for a few seconds, the liquid is poured into small cups, without refining or straining. Persons unaccustomed to this way of making coffee find it unpalatable. Those who have overcome the first introduction prefer it to that made after the French fashion, whereby the aroma is lost or deteriorated. A well made cup of good Turkish coffee is indeed the most delectable beverage that can be well imagined, being grateful to the senses and refreshingly stimulant to the nerves. Those who have long resided in the East can alone estimate its merits."—WHITE's *Three Years in Constantinople*.

"The Turkish way of making coffee produces a very different result from that to which we are accustomed. A small conical saucepan, with a long handle, and calculated to hold about two table-spoonfuls of water, is the instrument used. The fresh roasted berry is pounded, not ground, and about a dessert-spoonful is put into the minute boiler; it is then nearly filled with water, and thrust among the embers; a few seconds suffice to make it boil, and the decoction, grounds and all, is poured into a small cup, which fits into a brass socket much like the cup of an acorn, and holding the china cup as that does the acorn itself. The Turks seem to drink this decoction boiling, and swallow the grounds with the liquid. We allow it to remain a minute, in order to leave the sediment at the bottom. It is always taken plain; sugar or cream would be thought to spoil it; and Europeans, after a little practice (longer, however, than we had), are said to prefer it to the clear infusion drunk in France. In every hut you will see these coffee-boilers suspended, and the means for pounding the roasted berry will be found at hand."—CHRISTMAS's *Shores and Islands of the Mediterranean.*

"A small vessel, containing about a wine-glass of water, is placed on the fire, and, when boiling, a teaspoonful of ground coffee is put into it, 'stirred up, and it is suffered to boil and 'bubble' a few seconds longer, when it is poured (grounds and all) into a cup about the size of an egg-shell, encased in gold or silver filigree-work, to protect the finger from the heat; and the liquid, in its scalding, black, thick, and troubled state, is imbibed with the greatest relish. Like smoking, it must be quite an acquired taste."—MAXWELL's *Shores of the Mediterranean.*

CHICORY.

THE term chicory is an Anglicised French word, the original being chicorée. The plant is known to botanists by the name of *Cichorium Intybus*, and belongs to the natural order Compositæ, tribe Cichoreæ. It is an indigenous plant with a perennial root, better known probably to most readers by its English appellation of wild succory. The root is spindle-shaped, with a single or double head; externally it is whitish or greenish yellow; internally, whitish, fleshy, and milky. The roots grown in this country are smaller, and more woody or fibrous than those which are imported from the Continent.

The cultivation and consumption of chicory have now attained a very great importance, not only on the Continent, but also in the United Kingdom. Dating its extended use chiefly from the system pursued by the first Napoleon to substitute home-grown for colonial products, it has gradually become approved and popularised for a beverage, either used alone or more generally mixed with coffee, in numerous countries, where it can be sold far under the price of even the lowest grade coffees.

The manufacture of a factitious coffee from roasted chicory-root would seem to have originated in Holland, where it has been used for more than a century. It remained a secret until 1801, when it was introduced into France by M. Orban of Liége, and M. Giraud of Homing, a short distance from Valenciennes. This root is not superior to many others which possess sweet and mucous principles, but of all the plants which have been proposed as substitutes for coffee, and which, when roasted and steeped in boiling water, yield an infusion resembling the berry, it is the only one which has maintained its ground. The French, not satisfied with chicory, have recently introduced acorn coffee and roasted beetroot. The beet, it is asserted, besides communicating its hygienic qualities, also helps to sweeten the beverage. This new coffee is called "café de betterave," as the old was called "café chicorée." These distinctions will soon become as puzzling as those in America, which led the Irish waiter to ask if the gentleman would have coffee-tay or tay-tay.

Mr. George Phillips, when giving evidence before Mr. Scholefield's Parliamentary Committee on Adulteration, in 1855, stated that, prior to the year 1832, little was heard of the use of chicory in this country, but in the subsequent three years its use had gradually so increased that the Board of Inland Revenue was obliged to take steps against the sale. "I have no doubt (he adds), from my own experience, that a very large bulk of the public prefer the mixture. That, however, is a matter of taste. The trade contend that good coffee, mixed with one-eighth part of chicory, and sold at a moderate price, makes a better beverage than ordinary coffee would do at the same price, and the great mass of the public prefer it. Chicory sold as coffee yields a certain profit, but probably it equalises itself in the general competition of trade. There

is a large quantity of chicory sold by itself, and drank as a
beverage in the neighbourhood of Manchester and Liverpool.
I believe the price of a pound of the cheapest kind of coffee,
purchased by the bulk of the poor people, and a pound of the
mixture, is about the same. The trade say, when we use a
portion of chicory we use a better coffee. I do not know the
fact of my own knowledge. Whether the coffee sold in
mixtures is of a superior quality to that sold as a pure article
would be very difficult to ascertain; it depends upon the
question of taste and aroma. The chicory itself is not always
pure."

On the first introduction of chicory into Great Britain a
nominal duty of 20 per cent. was levied on it, which, owing
to the representations of the coffee-planters, was afterwards
increased to the same rate as that then payable on British
plantation coffee. The high duty thus levied on foreign-
grown chicory soon led to its cultivation in England, but so
little was known of the plant that the farmers required the
rent to be paid in advance for the use of their land. In the
autumn of 1853 we find chicory grown in Kent, Surrey, and
Essex, where the article was prepared, and met with a large
sale. With the increasing demand for the root, its culture spread to Bedford, Norfolk, Suffolk, Cambridgeshire,
Leicestershire, Cheshire, and Yorkshire. At first the price
realised was as high as 50l. per ton ground, and 20l. per ton
in the root. But as the growth extended the price receded.
The admission, duty free, of foreign-grown chicory, in 1854,
led to the abandonment of much of the home culture.

In 1842, Mr. McCulloch assumed the growth and consumption of chicory in the United Kingdom to be 6¾ million
pounds; in 1850, from careful inquiries I instituted, I estimated the consumption then to be double that amount. Mr.
Braithwaite Poole, in his "Statistics of Commerce," published in 1852, rated the actual production of chicory-root,

made into powder in England and Guernsey, then as high as
14,000 tons, worth, at 22*l*. per ton, 308,000*l*. The gradually
increasing imports of foreign-grown replaces much formerly
produced at home, but the changes in legislative enactments
have much interfered with the consumption of chicory here,
and hence the import is not so remunerative. From 1856 to
1859 the imports of foreign chicory in the root rose from
81,721 cwts. to 267,000 cwts., but there has since been a
gradual decline to 45,563 cwts. in 1862. The value has
ranged from 6s. to 10s. 6d. per cwt.

The largest quantity comes from Belgium, the next from
Holland, and a little from Hamburg and other quarters.
There are also some considerable imports of roasted and
ground chicory, which is chiefly re-exported ; 76,206 lbs. of
chicory-powder were imported in 1862.

Roasted and reduced to powder chicory is the most
universal substitute for coffee in the chief continental
countries, especially in France, Switzerland, Germany,
Belgium, Holland, Denmark, Russia, and other Northern
States. In Germany, the ground chicory is made up into
cakes, and sold in that form. Denmark and the Duchies
consume about 3,000,000 lbs. annually. A few years ago
the annual import of chicory-root into Hamburg was
24,600 cwts., and of ground chicory and other coffee substitutes 13,000 cwts.

Belgium exports 6,000,000 or 7,000,000 lbs. yearly. The
quantity of the dried root consumed in France is about
16,000,000 lbs. a year. Formerly they were able to export
1,000,000 or 2,000,000 lbs., but now enough is not produced for home consumption. In 1860, about 10,000,000 lbs.
of chicory-root was imported into France, chiefly from Belgium, and about 660,000 lbs. of chicory in powder was exported, chiefly to Algeria. Till within a few years the cultivation was carried on principally near Valenciennes, but

lately manufactories have sprung up in several localities, especially at Arras, Cambray, Lille, Paris, Senlis, in Normandy, Brittany, &c. In some parts of Germany the women are becoming regular chicory-topers, making of it an important part of their daily sustenance.

SECTION II.

CULTIVATION, HARVESTING, AND PREPARATION FOR MARKET.

THERE are many varieties of this plant, the greater part of which have blue flowers; some are white, and others red. In Brunswick they only grow the broad-leaved, or native kind, or the small-leaved, which has long roots, and is a native of Magdeburg. The former is, however, preferred, on account of its being the richest. In Altona they grow a medium variety, which has neither very narrow nor very broad leaves. The plant thrives in all soils that will grow carrots; indeed, the mode of cultivating one is much like that of the other. The roots seem, however, to grow best upon a loamy soil, with a clayey subsoil, dry, deep, and rich. It very seldom thrives in heavy clay land, and never in sand or wet land. It requires much manure. In preparing the land deep ploughing is recommended; but, unless the soil is very deep, it is probable that subsoil ploughing will answer better. The surface must be well worked; indeed, it cannot be reduced to too fine a mould.

As the plants are a long time in coming up, generally five or six weeks from the time of sowing the seed, it is necessary that the land should be very clean, or the weeds (particularly chickweed) are liable to overtop and smother the young plants. The time of sowing varies in different districts; in the midland and eastern counties of England, the second or third week in May is considered best, for if sown earlier, many of the plants will run to seed, in which case they are called "runners" or "trumpeters," and must be carefully dug out and destroyed when the time for taking up has arrived, because if allowed to become mixed with the bulk, they will spoil the sample. The best crops have been obtained when the seed has been sown broadcast; but the preference is usually given to drilling, the crop being more easily hoed and cleansed. The rows are generally from 9 to 12 inches apart, and about 3 or 4 lbs. of seed per acre is the quantity used.

Most of the cultivators of chicory single out the plants so as to leave spaces between them in the rows, each about 6 or 8 inches long; but there are many who do not do this, fancying that four or five small plants produce more weight of root than one large plant; the expediency of this, however, is very questionable, as it does not allow of the land being nearly so well cleaned as when the practice of singling is adopted.

In October or November, the work of taking up the roots may be commenced, and continued during the winter (if the crop cannot be previously secured), until it is finished. Although the roots penetrate a long way downwards, they become too thin below 14 or 15 inches to be useful, and the utmost care is also required in order to get up that portion of the root which will prove profitable.

In some cases chicory has been ploughed up, about 12 inches deep, with a strong cast-iron plough drawn by six horses, having men to fork each furrow to pieces with common potato-forks before a second furrow is ploughed upon it, and women and children following to pick up the roots and cut off the tops.

But the best method is found to be that of digging up the roots with double-pronged strongly-made iron forks, the blades being about 14 inches in length, and each fork, with shaft and handle complete, weighing about 8 lbs.

The plan of ploughing is liable to bring too much of the

subsoil to the surface, and costs quite as much, if not more, than digging.

The advantage which is looked for in ploughing, is to ensure getting the roots up from a greater depth than can be done by digging, as a great number break off about 8 or 9 inches long, unless a boy is employed to assist the diggers, and is very careful to pull the top at the precise time that the man presses the root upward with his fork.

When dug, the tops should be neatly cut off, and the roots conveyed to the washing-house to be cleaned. Sometimes they are earthed in pits, but, generally speaking, they are taken to the washing-house immediately after being dug up.

In the former case, on the Continent, the roots, with the leaves cut off, are thrown, in heaps of from four to six feet in length, width, and height, on the surface of the ground; some straw and then some earth are put around. But generally the growers deliver the roots to the manufacturers from the latter end of August to November, by whom they are immediately dried.

The root is from 2 to 4 inches thick, 3 to 7 inches long, and occasionally, in a good soil, 3 lbs. in weight. In Brunswick they obtain from 4 to 6 tons of root per Brunswick acre.

The weight of the crop depends entirely upon the richness or poverty of the soil, the tillage and manure it has received, and other circumstances. The fault in England is the striving to grow as heavy a crop as possible, to the very great detriment of the quality of the root for powder.

In Brunswick the price of the root in the original state varies from 20s. to 40s. per ton, according as the crops have been good or bad, and an acre will realise from 5l. to 7l. The cost for cultivation is from 3l. 15s. to 4l. 10s.; 1½ to 2 tons is about an average crop.

Mr. William Strickney, who has grown and prepared chicory for the manufacturer to a very great extent, on a large farm near Hull, estimates the expense of the cultivation of chicory there at 4*l*. 5s. 6d. per acre, and if we add to this 2*l*. 10s. for rent, manure, &c., it gives 6*l*. 15s. 6d. The produce on suitable land he states to be from 8 to 12 tons per acre, and it requires 4 tons of green root to make 1 ton of dried. In the dried state the root is worth from 12*l*. to 24*l*. per ton. Take 10 tons per acre, at 2*l*. 10s. per ton, and this would leave a profit per acre of 18*l*. 4s. 6d.

Another competent agricultural authority states that the price of $2\frac{1}{4}$ tons of dry root for the acre, at 12*l*. per ton, would be 27*l*.; deducting 7*l*. for rent, labour, and other expenses, this would leave a profit of 20*l*. per acre.

The roots are cut into small pieces of about half-inch or three-quarter inch lengths by a turnip-cutter, or by hand, the object being to have the pieces of as uniform a size as possible. The slices are then dried in a kiln : this process wasting the chicory from 75 to 80 per cent. It is then marketable, and is usually sold to the drysalters and grocers, who roast and grind it as they do coffee. In the ground state it may be kept for years, but it soon cakes. The roasted root is emptied into iron vessels, and, after cooling, is crushed in vertical stone mills, or between iron cylinders.

The dried roots cut are roasted in this country like coffee. The loss during roasting is from 25 to 30 per cent. The roasters generally introduce into the roasting machine about 2 lbs. of lard for every cwt. of chicory. Some say this is to give the chicory a better face, others state that it renders the powder less hygrometric. Inferior kinds of chicory are alleged to be coloured with Venetian red.

Chicory is occasionally adulterated with roasted pulse (called Hambro' powder), damaged corn, and coffee husks ("coffee flights," as they are technically termed). We have also heard of parsnips having been roasted, ground, and mixed with chicory. Dr. Hassall gives a long list of other substances which have been found as adulterants of coffee.

Treacle is sometimes introduced into fictitious chicory, to give the caramel or saccharine odour possessed by real chicory.

Dr. Hassall says the roasted chicory root yields from 45 to 65 per cent. of soluble extractive. Its solution in water is acid, and it does not possess the peculiar bitter taste of the raw root; but the taste of the liquid is more like that of burnt sugar. The copper test shows the presence of from 10 to 13 per cent. of sugar.

SECTION III.

STRUCTURE AND CHEMICAL COMPOSITION.

THE following analysis represents the per-centage composition of chicory root in its different conditions:

	Raw root.	Kiln dried.
Hygroscopic moisture	77·0	15·0
Gummy matter (like pectine)	7·5	20·8
Glucose, or grape sugar	1·1	10·5
Bitter extractive	4·0	19·3
Fatty matter	0·6	1·9
Cellulose, inuline, and woody matter	9·0	29·5
Ash	0·8	3·0
	100·0	100·0

The composition of the roasted root was as follows:

	1st species.	2nd species.
Hygroscopic moisture	14·5	12·8
Gummy matter	9·5	14·9
Glucose	12·2	10·4
Matter like burnt sugar	29·1	24·4
Fatty matter	2·0	2·2
Brown or burnt woody matter	28·4	28·5
Ash	4·3	6·8
	100·0	100·0

Dr. Hassall gives the following results of trials instituted to determine the effect of chicory on the human frame.

Three persons partook of a chicory breakfast. The infusion was dark-coloured, thick, destitute of the agreeable and refreshing aroma so characteristic of coffee, and was of a bitter taste.

Each individual experienced for some time after drinking this infusion a sensation of heaviness, drowsiness, a feeling of weight at the stomach, and great indisposition to exertion; in two headache set in, and in the third the bowels were relaxed. In second and third trials of the chicory breakfast the same feelings of drowsiness, weight of the stomach, and want of energy were experienced, but no headache or diarrhœa. Several other trials were subsequently made, with nearly similar results. But chicory, it will be said, is seldom taken alone in this country, and when mixed with coffee these effects are not produced.

Two persons partook, for a considerable period, twice a day, of an article denominated coffee, costing 1s. 6d. a pound, and largely adulterated with chicory; during nearly the whole of this time they both suffered more or less from diarrhœa.

From the results of these trials, therefore, we are warranted in concluding that at least some doubt is attached to the assertion of the wholesome properties of chicory-root as an article of diet.

Several characters, sufficiently simple and easily recognised for general application, have been indicated in different works for detecting the addition of roasted chicory to coffee in the roasted and ground state, but the application of chemical reagents for detecting the presence of the colouring matter of roasted chicory, when added to infusion of coffee, has not yet proved successful.

The brownish-yellow colouring matter which is developed in chicory-root by the process of roasting, when dissolved in water by infusion or decoction, retains its colour, or becomes a little deeper by the action of persalts of iron, without giving rise to any precipitation.

The brown colouring matter of roasted coffee, on the other hand, acquires, from the same reagent, a green colour, and a brownish-green flocculent precipitate is formed. These two different reactions may be applied, not only for distinguishing the pure infusion of coffee and of chicory, but also those which contain a mixture of the soluble principles of the two alimentary substances.

Infusion of pure coffee acquires a green colour, more or less intense, on the addition of some drops of persulphate of iron.

Infusion of pure chicory, under similar circumstances, retains its brownish-yellow colour, which becomes more intense, and acquires a slight greenish tint.

A mixture of the two infusions, containing one-half, a fourth, or a fifth of its volume of infusion of chicory, may be recognised by its brownish-yellow colour, which remains after the deposition of the precipitate produced by the salt of iron, together with part of the colouring matter of the coffee. This separation may be expedited by rendering the coloured liquor slightly alkaline by the addition of a small quantity of weak solution of ammonia, and allowing it to stand in tubes closed at one end. The supernatant liquor, after the precipitate has deposited, will possess a brownish-yellow tint by refracted light, which will be deeper in proportion to the quantity of chicory present.

If the experiment be first made with infusion of pure coffee of a certain strength, and afterwards with additions of known quantities of chicory, keeping these for comparison, the quantity of chicory in a mixed sample may be thus determined.[*]

A simple means of detecting the chicory in ground coffee is as follows :

Throw about a tea-spoonful of the suspected coffee in a wine-glass of water, and stir the mixture with a spoon. If

[*] M. Lassaigne, in "Journal de Chimie Médicale."

Plate 10.

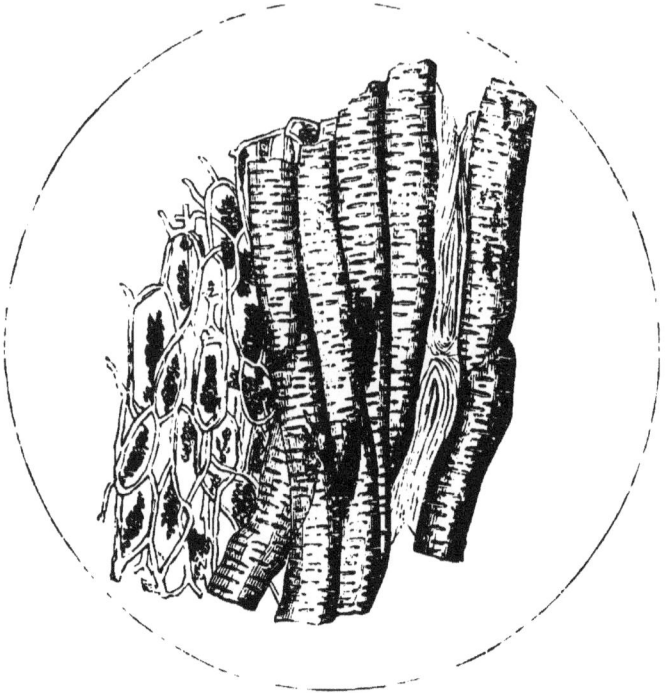

Fragment of Roasted Chicory.

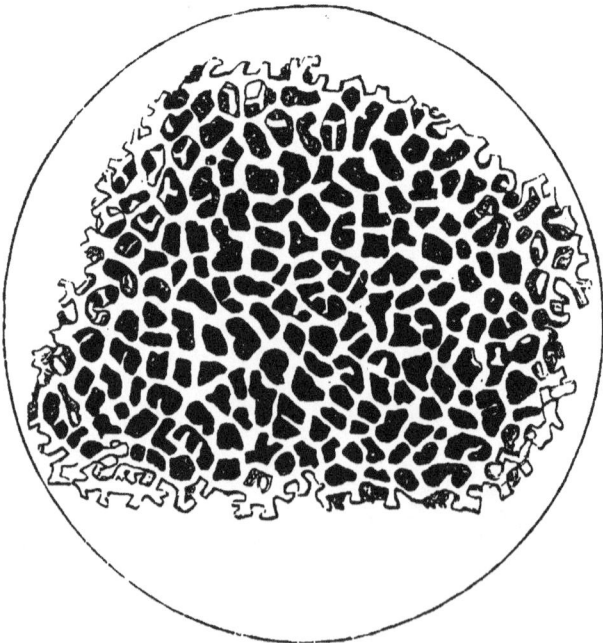

A Fragment of Roasted Coffee, being magnified 140 diameters.

the coffee be pure, it for the most part floats, becomes very slowly moistened, even when shaken up with the water, and communicates scarcely any colour to the liquid; very gradually it imbibes water; the liquid acquires a very pale sherry tint; and at the end of several hours the greater part of the powder is found to have fallen to the bottom of the glass. If, however, it be chicorised, the presence of chicory (genuine or spurious) will be readily detected, by a portion of the suspected powder rapidly sinking and communicating to the liquid a reddish-brown tint, which will be more or less deep according to the amount of chicory present.

If the coffee be adulterated with what is called Hambro' powder (roasted and ground peas, &c., coloured with Venetian red) or roasted corn, we have a further test in iodine, which communicates a purplish or bluish-red tint to the water to which either of these substances has been added. The preceding test is sufficiently delicate and valuable, in all ordinary cases, for detecting chicory in coffee; but to those familiar with microscopic investigations, the microscope furnishes another mode of proceeding: fragments of dotted ducts being found in chicory, but not in pure coffee. They are not met with, however, in great abundance; and some patience and care, therefore, are requisite in searching for them. The starch grains of Hambro' powder are readily detected by the microscope, as also the blackening effect of a solution of iodine on them.

Plate 10 represents the structure and character of genuine ground roasted coffee, and of a fragment of roasted chicory-root, showing the dotted or interrupted spiral vessels which pass in bundles through the central parts of the root, magnified 140 diameters; copied, by permission, from Dr. Hassall's work on " Food and its Adulterations."

In the raw chicory-root three parts or structures may be distinguished with facility, cells, dotted vessels, and vessels

I

ŏf the latex. These vessels afford useful means of distinguishing chicory from some other roots employed in the adulteration of coffee. The chief part of the root is made up of little utricles or cells. These are generally of a rounded form, but sometimes they are narrow and elongated. The former occur when the pressure is least and the root soft, the latter in the neighbourhood of the vessels.*

There are four characters by which adulterated chicory may be distinguished from the genuine.

1st. It yields to cold water a much whiter colour. In using this test it is necessary to have a sample of genuine chicory for comparison.

2ndly. A decoction of chicory containing either roasted grain or pulse, yields when cold a purplish or bluish-black colour, with a solution of iodine; whereas a corresponding decoction of genuine chicory is merely coloured brown by iodine.

3rdly. The microscope detects in adulterated chicory the torrefied starch grains of either corn or pulse. That they are starch grains is shown by the action of a solution of iodine, which blackens them.

4thly. The odour and flavour will sometimes detect adulterations.

Roasted and ground chicory attracts water from the air, and thereby increases in weight and becomes clammy. The grinders are accustomed to return as much by weight of ground chicory as they receive of the unground root, for the loss which the root suffers by grinding is more than compensated by the absorption of water from the air.

* " Food and its Adulterations."

THE END.

C. WHITING, BEAUFORT HOUSE, STRAND.